Nutrient Content
of
Food Portions

JILL DAVIES

Department of Hospitality, Food and Product Management
South Bank Polytechnic, London

and

JOHN DICKERSON

formerly Professor of Human Nutrition
School of Biological Sciences
University of Surrey, Guildford, Surrey

The Royal Society of Chemistry

Published by: The Royal Society of Chemistry
Thomas Graham House
Science Park
Milton Road
Cambridge CB4 4WF
UK

Tel.: (0223) 420066   Telex: 818293

Production: Ian Unwin
Cover design: John Tanton

ISBN 0-85186-426-0

Orders should be addressed to:
The Royal Society of Chemistry
Distribution Centre
Letchworth
Herts. SG6 1HN
UK

Xerox Ventura Publisher™ output photocomposed by Goodfellow & Egan Phototypesetting Ltd, Cambridge

Printed in the United Kingdom by
The Bath Press, Lower Bristol Road, Bath

CONTENTS

FOOD SECTIONS in the main tables

FOREWORD

by Dr Elsie M Widdowson CBE FRS

Fifty-one years ago the first edition of our 'Chemical Composition of Foods' was published. This contained values for the amounts of water, protein, fat, carbohydrate, energy and minerals per 100 grams and per ounce of 541 foods. Both raw and cooked foods were included, and also about a hundred made-up dishes. A second edition appeared in 1946, which was extended to contain values for wartime foods such as Household milk, dried eggs and National flour. Values for economical cooked dishes were added. In 1960 a third edition was produced which for the first time contained values for the amounts of vitamins in foods. The fourth, current edition was published in 1978. It was extended and revised by Miss Alison Paul and Professor David Southgate, both colleagues working in our Department. The ounce section was removed, and all values were expressed per 100 grams of food.

Now Professor John Dickerson, another colleague and friend of many years, and Dr Jill Davies have added a further dimension to the way the values are presented, by discarding weighing machines and expressing the amounts of nutrients per average portion. Great care was taken to determine the size of the 'average portion', but this presents difficulties because none of us is 'average', and one of the remarkable characteristics of individuals is the wide variation in the amounts of food and hence of energy and nutrients that individuals take. However, this can be allowed for by allotting the big eater, say, two average portions and the small eater half a portion. These new tables contain values for many foods that have come on the market in recent years, for a variety of cooked dishes and for foods favoured by immigrant populations now living in the UK. I believe they will fulfil a need that has not hitherto been met and I wish the book every success.

Elsie M Widdowson

February 1991

INTRODUCTION

1.   Why food portions?

The nutrients required for the healthy working of our bodies are derived from food and drink. Since no single food contains adequate amounts of nutrients, we must generally consume a variety of foods. If the variety is restricted, much more care is needed in the choice of foods in order to maintain a nutritionally adequate diet. The variety of foods available in western societies is quite enormous, for besides what may be considered the 'natural' foods – milk, eggs, meats, vegetables and fruits, there is an increasing variety of 'prepared', 'convenience' and 'fast' foods. The choice of foods is very much an individual matter and is influenced by a complex interplay of factors – race, culture, religion, ethical considerations, age and physiological condition, quite apart from life-style and economics.

The nutritional adequacy of a diet is determined by the nature and amounts of the foods consumed, and their nutrient content. The nutritional content of a diet is, therefore, usually assessed by determining the amounts of the various foods consumed and calculating, with the aid of tables of food composition, the amounts of the various nutrients which the foods contain and so deriving values for the total energy, protein, fibre and other nutrients contained in the diet. Dietary adequacy can then be assessed, if desired, by comparing these values with the 'Recommended Daily Amounts' (RDA's) published periodically in the U.K. and other countries, principally the United States (U.S.).

Assessing the nutrient content of a diet can be a tedious and demanding exercise depending on the interest, conscientious application and commitment of the subjects and the time of the investigator. For some purposes, the precision to be gained by such measurements is unnecessary and their demanding nature often results in the omission of an otherwise valuable piece of information. This may be true in education, in monitoring one's personal diet and in the assessment of the adequacy of the nutritional intakes of the sick.

Nutrition is an important component of Home Economics courses in schools and colleges and the assessment of nutrient intake and dietary adequacy a necessary part of the associated practical work. It might be rather more important for students to be able to gain a knowledge of a broad range of nutrient intakes derived from a variety of diets than to make precise measurements on a few. With reference to an individual's diet, it is more valuable to be able to do an assessment quickly and easily with small demands on skill and knowledge. Furthermore, in hospitals it may well be sufficient to know whether a patient is consuming half or less of his probable requirement of various nutrients (Duthie, 1988). Food choice is now often a patient's own responsibility and it is desirable that the nutritional consequences of poor food choice can be identified quickly.

On the one hand we recognise that precise dietary assessment is a demanding and time-consuming exercise. On the other, there is a recognised need for such assessments but with rather less precision. There are two ways in which the process can be greatly simplified. The need for foods to be weighed can be obviated by the use of previously determined average sizes for portions of commonly consumed foods. Then, by having available tables of food composition which contain the pre-calculated amounts of nutrients in these portions, the time required for calculation of the nutrient content of the diet can be reduced to a minimum. Thus, no scales are needed, only records of the nature and frequency of foods consumed, for example food frequency questionnaires and food record diaries. Then, with the aid of the tables in this book, food portions can be converted easily into amounts of nutrients and a simple calculator is all that is needed to summate the nutrient content of the diet. Whatever method is used for the assessment of nutrient intake, it must be remembered that diets are variable from day to day. Therefore the greater the number of days involved, the more reliable will be the estimate.

2. Determination of portion sizes

The suggested use of average food portion sizes to replace the precise weighing of food items is not new. It is a much more practical approach to dietary analysis, reducing the time involved. The weighing of food tends to bias the choice of foods and the quantity eaten. Moreover, some foods are already packaged at point of purchase as 'portions' and others occur naturally in this way. As examples of the former we could consider frozen 'convenience' fish and meats and of the latter, eggs and apples. To work with tables of food composition based on unit weight, say 100g, is unrealistic in both cases.

Tables of average portion sizes have been published (e.g. Bingham & Day, 1987; Crawley, 1988). Clearly, portion sizes of commonly used foods can be determined in different ways. One can measure them prospectively in a planned manner after agreeing the method to be adopted, or values can be extracted retrospectively from the records obtained during dietary surveys in the community (e.g. Bingham et al., 1981). Crawley (1988) obtained hers from the records which were obtained during the Ministry of Agriculture, Fisheries and Food (MAFF) Household

Survey. Because some foods are consumed much more frequently than others, the analysis of retrospective data results in the use of varying numbers of observations to construct one 'average' portion size. Extremes of size also tend to be recorded in dietary surveys. It could be said that the mean of very small and very large portions represents an 'average' portion. It is possible both to justify and to criticise values obtained in this way. The whole question of food portions is highly controversial and a definition of what constitutes a 'standard' or 'average' is unlikely to be generally acceptable. **The food portions used in these tables are considered as estimates and should be interpreted as such.**

The portion sizes used for these tables were obtained prospectively and intentionally as described previously (Davies & Dickerson, 1989). All foods were purchased and either used as such or made up according to standard recipes quoted in Tables of Food Composition. To estimate portion size three different people were asked independently to serve what they considered to be a small, medium and large portion. Appropriate serving vessels, for example dinner plates, soup bowls or mugs, were used for this procedure so that the size of the portion was judged in its 'normal serving environment'. All weighings were recorded on scales accurate to the nearest gram. Nine weights – three small, three medium and three large – were obtained for each food item; the small and large portions were used to validate the medium-sized portion. The mean value of the medium-size portion was used in the tables. When foods purchased were already in the form of a unit, such as one pork chop, one brown roll, one jam doughnut, three different brands were obtained and the mean value taken as the weight of the item. For items such as biscuits, three different people were asked to judge small, medium and large portions in order to assess whether one or more units constituted a medium-sized portion. For fruit (apples, pears, oranges, etc.) the weight of three small, three medium and three large examples were obtained and the mean value of the medium sized fruit used for the tables. For alcoholic drinks, the portions quoted are based on the measures used in public houses.

The nutrient content of the food portions was calculated from values in the standard U.K. Tables of Food Composition (Paul & Southgate, 1978), and additional data on cooked dishes (Paul & Southgate, 1979; Wiles et al.,1980). The U.K. Tables have more recently been revised and expanded in the supplements to 'The Composition of Foods': *Immigrant Foods* (Tan et al., 1985), *Cereal and Cereal Products* (Holland et al., 1988) and *Milk Products and Eggs* (Holland et al., 1989); these sources have been used as appropriate.

3.   *The composition of foods*

Most of the foods consumed by man are complex mixtures of organic substances (fats, carbohydrates, proteins and vitamins), minerals and water. The composition of foods varies with their source, whether they are derived from animals or plants.

Generally, foods of animal origin contain more protein and fat and smaller amounts of carbohydrate, whereas foods of plant origin contain larger concentrations of complex carbohydrates, such as starch, and smaller amounts of protein and fat. Plant foods also contain varying amounts of substances that are collectively described as 'dietary fibre'. These substances are not digested in the small intestine and are therefore not absorbed and should not be considered as nutrients. Nevertheless they have very important effects on bowel function and because of this may affect the absorption of nutrients.

Different foods vary considerably in their vitamin and mineral content. Foods of animal origin with their higher fat content tend to contain more of the fat soluble vitamins whereas the water soluble precursors of these are found in plants. The highest concentrations of fat soluble vitamins are found in organ meats (e.g. liver and kidney) in which they are stored in the live animal. Vitamin A is a good example of what we are talking about. Retinol, or pre-formed vitamin A, is found in highest concentrations in liver whereas its precursor, β-carotene, is found in leaves and coloured vegetables and fruits. An exception to this general comment is vitamin E (or α-tocopherol) which is not found in high concentrations in animal foods but for which wheat germ, or wheat germ oil, are good sources. In contrast, the water soluble vitamins are found in rather higher concentrations in plant foods. Vitamin B_2 (riboflavin) is an exception to this generalisation, for milk is the most important source in the diet of people in the U.K.

Minerals are found in both animal and plant foods. However, in plant foods the amounts present are not a reliable guide to their nutritional value. This is because a proportion of the particular mineral may be tightly bound to another food constituent in such a way that it is unavailable for absorption. Thus, what is known as the 'bioavailability' of minerals such as calcium and iron must be considered when assessing the nutritional value of the mineral content of a diet.

In addition to the nature of the food, its composition will vary according to the environment in which the animal or plant was grown – rate of growth, exposure to sunlight and the mineral content of the soil may all have an effect. Added to this may be the effects of harvesting, preservation, storage and processing (including cooking, and indeed different methods of cooking). When we consider preservation and processing there are changes in a food composition that occur because of the addition of one or more of a possible hundreds of substances known as 'food additives'. These may be added by the food manufacturer for a variety of reasons – as preservatives, as colorants, to change the texture, or as flavours. The presence and nature of these in foods manufactured or sold in the U.K. is now indicated on labels by 'E' numbers.

Table 1   *Recommended Daily Amounts of Energy and Nutrients (D.H.S.S., 1979)*

Age (years) and occupational category	Energy (Kcal)	Protein (g)	Fat-soluble vitamins		Water-soluble vitamins				Minerals	
			A (RE; μg)[1]	D (μg)	B1 (mg)	B2 (mg)	B3 (NE; mg)[2]	C (mg)	Ca (mg)	Fe (mg)
Infants under 1	950	25	450	7.5	0.3	0.4	5	20	600	6
Children 5-6	1710	43	300	a	0.7	0.9	10	20	600	6
Boys 12-14	2640	66	725	a	1.1	1.4	16	25	700	12
Girls 12-14	2150	53	725	a	0.9	1.4	16	25	700	12
Men 18-34 Sedentary	2510	63	750	a	1.0	1.6	18	30	500	10
Moderately active	2900	72	750	a	1.2	1.6	18	30	500	10
Very active	3350	84	750	a	1.3	1.6	18	30	500	10
35-64 Sedentary	2400	60	750	a	1.0	1.6	18	30	500	10
Moderately active	2750	69	750	a	1.1	1.6	18	30	500	10
Very active	3350	84	750	a	1.3	1.6	18	30	500	10
65-74 Sedentary	2400	60	750	a	1.0	1.6	18	30	500	10
75 + Sedentary	2150	54	750	a	0.9	1.6	18	30	500	10
Women 18-54 Most occupations	2150	54	750	a	0.9	1.3	15	30	500	12[c]
Very active	2500	62	750	a	1.0	1.3	15	30	500	12[c]
55-74 Sedentary	1900	47	750	a	0.8	1.3	15	30	500	10
75 + Sedentary	1680	42	750	a	0.7	1.3	15	30	500	10
Pregnant	2400	60	750	10	1.0	1.6	18	60	1200[b]	13
Lactating	2750	69	1200	10	1.1	1.8	21	60	1200	15

[1] RE = Retinol Equivalents: 1 Retinol Equivalent = 1μg retinol or 6μg β-carotene
[2] NE = Nicotinic acid Equivalents; 1Nicotinic acid Equivalent = 1mg nicotinic acid or 60mg of dietary tryptophan
a For children and adults who have sufficient exposure to sunlight no dietary sources of vitamin D may be necessary, but in winter children and adults should have 10μg daily by way of supplements. These supplements may also be required by adults who have inadequate exposure to sunlight, e.g. the housebound.
b Extra calcium needed in the 3rd trimester, i.e. 6th to 9th month of pregnancy.
c This amount of iron may not be sufficient for 10% of women who have large menstrual losses.

Some nutrients are also added, as required by law, to certain foods where it is desired for health reasons to protect people from deficiencies. Vitamins, iron and calcium have been particularly of concern. The nutritional value of some foods may be changed by the use of nutrients in their manufacture. Thus the food industry uses large amounts of vitamin C because of its reducing (or anti-oxidant) properties. Sugar is added to some foods such as jams as a preservative whilst it is added to others as a flavouring agent.

Against this background it will be appreciated that the analysis of foods is a specialised and somewhat daunting task. It must be a continuing task in an attempt to keep pace with increasing variations in the availability of foods. Apart from this, samples must be chosen with care in order to make them as representative as possible. Analytical methods must be carefully controlled. Made-up foods must be made according to standard recipes. Hundreds of foods have been analysed by the Laboratory of the Government Chemist, food manufacturers and research workers. The classic source book for food composition data in the U.K. has been McCance and Widdowson's *The Composition of Foods*, the fourth edition of which was published in 1978 (Paul and Southgate, 1978). However, it has been recognised that information about the composition of foods is an important national resource and moreover, one which should be kept up-to-date. The Royal Society of Chemistry, in collaboration with the Ministry of Agriculture, Fisheries and Food (MAFF), is now maintaining this database and its resources were made available to us to construct the tables presented in this book.

4. Selection of nutrients.

Practical usefulness has been a major consideration in the production of these tables based on portion sizes. For groups of healthy people the yard-stick for comparison of the nutrient content of diets is the table of Recommended Daily Amounts (DHSS, 1979) (Table 1). We have therefore chosen to include the nutrients given in these tables and to include them in a form in which they can be compared directly with the RDA's. Thus, values for vitamin A are quoted as 'retinol equivalents' to make allowance for the contribution of β-carotene, and those for vitamin B_3 as 'niacin equivalents' making allowance for the contribution of niacin formed from tryptophan. Values for vitamin D have been excluded because there is no RDA for the bulk of the population. Values are given only for young children and the housebound elderly. For the remainder of the population it is considered that there is not a dietary requirement. Exposure of the skin to sunlight during the summer results in the formation of sufficient vitamin D (cholecalciferol) from 7-dehydrocholesterol to last through the year.

6

Table 2 *Dietary Goals according to NACNE[1] and COMA[2]*

NACNE (1983) **BODY WEIGHT** Adjust types and amount of food eaten and increase exercise output to maintain optimal limits of weight for height.

FAT Reduce total fat to 30% energy intake.

CARBOHYDRATE Increase intake of food containing complex carbohydrates, e.g. starch.

PROTEIN Eat more vegetable protein at the expense of animal protein.

FIBRE Increase fibre to 30g per head per day.

COMA (1984) **BODY WEIGHT** Adjust food intake in relation to physical activity until weight is within acceptable range.

FAT Reduce total fat to 35% energy intake.

SATURATED FAT Reduce saturated fat to 15% energy intake.

PROTEIN No specific recommendation.

FIBRE Compensate for reduced fat with increased fibre rich carbohydrate foods.

[1]National Advisory Committee on Nutrition Education.
[2]Committee on Medical Aspects of Food Policy.

We have also been concerned that the tables should be useful for people wishing to adopt the guidelines for a healthy diet suggested by the National Advisory Committee on Nutrition Education (NACNE) and the Committee on Medical Aspects of food policy (COMA) (Table 2).

United States RDA's (Table 3) include additional vitamins and minerals (B_6, B_{12}, folates and vitamins D, E; Mg, Zn and I) and in certain assessments there may well be interest in these nutrients. Appendices 1-6 (pages 57-61) give some examples of foods that are useful sources of these nutrients.

5. The uses and limitations of food tables.

The authors of the best known U.K. food composition tables once wrote (Widdowson & McCance, 1943) "There are two schools of thought about food tables. One tends to regard the figures in them

Table 3 Recommended Daily Dietary Allowances of the Food and Nutrition Board, National Academy of Sciences National Research Council (Revised 1980)

	Age (years)	Protein (g)	Fat-soluble vitamins			Water-soluble vitamins							Minerals				
			A (RE; µg)[1]	D (µg)	E (TE; mg)[2]	B$_1$ (mg)	B$_2$ (mg)	B$_3$ (NE; mg)[3]	B$_6$ (mg)	Folicin (µg)[4]	B$_{12}$ (µg)	C (mg)	Ca (mg)	Mg (mg)	Fe (mg)	Zn (mg)	I (µg)
Infants	Under 1	Wt(kg)×2.1	410	10	3	0.4	0.5	7	0.5	38	1.0	35	450	60	13	4	45
Children	5-6	30	500	10	6	0.9	1.0	11	1.3	200	2.5	45	800	200	10	10	90
Boys	11-14	45	1000	10	8	1.4	1.6	18	1.8	400	3.0	50	1200	350	18	15	150
Men	19-22	56	1000	7.5	10	1.5	1.7	19	2.2	400	3.0	60	800	350	10	15	150
	23-50	56	1000	5	10	1.4	1.6	18	2.2	400	3.0	60	800	350	10	15	150
	51 +	56	1000	5	10	1.2	1.4	16	2.2	400	3.0	60	800	350	10	15	150
Girls	11-14	46	800	10	8	1.1	1.3	15	1.8	400	3.0	50	1200	300	18	15	150
Women	19-22	44	800	7.5	8	1.1	1.3	14	2.0	400	3.0	60	800	300	18	15	150
	19-22	44	800	5	8	1.0	1.2	13	2.0	400	3.0	60	800	300	18	15	150
	75 +	44	800	5	8	1.0	1.2	13	2.0	400	3.0	60	800	300	10	15	150
Pregnant		+30	+200	+5	+2	+0.4	+0.3	+2	+0.6	+400	+1.0	+20	+400	+150	a	+5	+25
Lactating		+20	+400	+5	+3	+0.5	+0.5	+5	+0.5	+100	+1.0	+40	+400	+150	a	+10	+50

[1] RE = Retinol Equivalents: 1 Retinol Equivalent = 1µg retinol or 6µg β-carotene
[2] TE = α-Tocopherol Equivalents.
[3] NE = Nicotinic acid Equivalents; 1Nicotinic acid Equivalent = 1mg nicotinic acid or 60mg of dietary tryptophan
[4] The folicin ('folate') allowances refer to dietary sources as determined by *Lactobacillus casei* assay after treatment with enzymes (conjugases) to make polyglutamyl forms of the vitamin available to the test organism.
[a] The increased requirement during pregancy cannot be met by the iron content of habitual American diets nor by the existing iron stores of many women; a supplement of 30-60mg of iron is recommended.
N.B. Phosphorus has been omitted from this table as deficiency of this element is unknown in man.

as having the accuracy of atomic weight determinations; the other dismisses them as valueless on the grounds that a foodstuff may be so modified by the soil, the season or its rate of growth that no figure can be a reliable guide to its composition. The truth, of course, lies somewhere between these points of view". It is unlikely that anyone will claim undue accuracy for tables based on average portion sizes. Clearly, they are at best approximate but, as stated above, they are particularly useful as a **dietary ready reckoner** for personal, community health and clinical purposes.

Individuals of any age or physiological condition may wish to ascertain the approximate energy, protein and fibre content of their diets. Parents are concerned about the intakes of their children. The pregnant and, particularly, the lactating woman is concerned that she is having sufficient but not overmuch food. Those having care of the elderly will also find it helpful to have some idea about dietary intakes. But very often the intakes of minor nutrients are ignored because their calculation using generally available food tables entails too much effort with a calculator. It is often the intakes of the minor nutrients that are the greatest cause for concern. It is so in each of the groups mentioned here. These tables make it possible for anyone to obtain quickly quite a detailed picture of the nutritional value of a diet.

Dietary assessments in groups of people in the community are expensive to carry out because they are so time-consuming. The population studied tends to be a self-selecting one because of the demands made on the subjects. By using portion sizes, and particularly tables giving the nutrient content of portions, the procedure for both investigator and subject is simplified. Dietary surveys cease to be the province solely of the expert and become a possibility for the home economist and market assessor. We need to know more about the effects on nutrient intake of greater dependency on convenience and fast foods in sections of the population. Possible target groups for new foods may be quickly and unobtrusively identified. Members of ethnic minorities may be more willing to take part in dietary studies if the recording of their diets is simplified.

For the nurse caring for a sick patient whose nutrition has been limited by their disease, for an individual suffering from a disease such as diabetes mellitus ('sugar diabetes'), or attempting to follow a diet aimed at lowering blood lipids (e.g. cholesterol), the availability of a means of easily assessing dietary intake can increase compliance, confidence and add considerably to dietary variations. In all such conditions it is insufficient to be able to assess major food constituents. Instructions given by dietitians may be either too precise as to be limiting or not sufficiently precise so that too much latitude is given to the patient (Lambert & Dickerson, 1989). Patients given nutritionally-based advice will be able, with simplified food tables of the kind presented in this book, to modify their own diets yet be confident that they are within prescribed guidelines. Failure to comply with dietary advice may be a common reason for the failure of dietary treatment.

Reference has been made earlier to the usefulness of the tables to those studying Home Economics. With the current interest in healthy eating, the tables would provide such students with a means of more easily planning healthy meals. Nowadays 'Home Economics' comes under a rather wider umbrella of 'Design and Technology' so the tables could be useful in the large and small scale planning and evaluation of diets.

6. Beyond the tables......

The tables given in this book give the nutritional content of selected foods. There are many more foods that might have been included. Some of these are packaged and the packages carry labels. The labels must, by law, carry certain information - the name of the food, net quantity, list of ingredients, minimum durability ('best before', 'sell by'), special conditions of storage, name and address of manufacturer, packer or distributor, place of origin and instructions for use. In addition, there is some nutritional information - ingredients, and nutritional information (nutrient content per serving, per 100g weight, or as a percentage of the RDA per serving). There may also be statements which make certain nutritional claims for the food (Davies & Dickerson, 1989). If they are read carefully, the information that they contain can be used in any dietary assessment. Thus, the absence of a food from our tables will not mean that it cannot be included in a diet.

Furthermore, as pointed out earlier, the whole food market, with its many and varied outlets, is a dynamic part of life in western societies. Quite apart from the indigenous development of new foods, the increased, and increasing facility for travel, whether for business or pleasure, leads to increasing variety in personal menus. For this reason it may well be necessary to revise our list of foods in a few years. Evidence of the likelihood of this is found in a recent supplement (Holland *et al.*, 1989) to the fourth edition of McCance and Widdowson's, *The Composition of Foods* (Paul & Southgate, 1978). The section of milk products and eggs in the 1978 edition contains 54 foods, the supplement 335.

Main Tables

KEY to conventions used in the tables

Nutrient abbreviations

Pro   Protein

Carb   Carbohydrate

Fib   Dietary fibre

Ca   Calcium

Fe   Iron

Alc   Alcohol

Vitamin names

A   Retinol equivalents; contibuted by retinol and carotenoids

B_1 Thiamin

B_2 Riboflavin

B_3 Nicotinic acid equivalents, consisting of available nicotinic acid and nicotinamide (collectively known as niacin) together with a contribution from tryptophan

C   Ascorbic acid

Nutrient values

0   None of the nutrient is present

Tr   Trace

N   There is no reliable information on the amount of nutrient present

Other abbreviations

(fl) oz   (fluid) ounce

lb   pound

pt   pint

TABLE FORMAT

Foods (including beverages) are listed alphabetically in sections, which are also in general arranged alphabetically. These food sections have been listed on the page (p. *iv*) following the Contents page. Alcoholic drinks are placed at the end of the tables (p. 53); for these the portion is expressed as a volume rather than by weight.

Each individual item has an identification number. The food and portion are described as necessary and the portion size is quantified. The nutrients covered in the main tables have been selected taking into account UK RDA's and NACNE healthy eating guidelines. The amounts of these nutrients given apply to the stated portion size rather than the 'per 100g' of conventional Food Composition Tables.

The alphabetical listing of foods in groups allows users to look up foods in the Tables conveniently, to review the alternative items covered and thus to select the 'right' food or the nearest equivalent. In some cases where foods are used in distinctly different amounts according to the situation, for example milk in tea and on breakfast cereal, they are appear in more than one group. However all food items have an identification number and can readily be found by using the index (p. 65). Foods usually consumed together have been listed individually whenever possible to allow for variations in food choice. For example a cup of tea may be taken black, with different types of milk, with a wedge of lemon, and sugar.

Additional description of food items is essential when the basic food name of the item is not self-explanatory. This aspect is particularly important when variations in ingredients occur such as the use of skimmed milk in place of whole milk in a given recipe. The portion weight may include inedible matter if this forms part of the serving, but parts of the food described detailed are considered eaten. For example, in 'Roast chicken, meat and skin' all 85g is consumed, but for 'Roast chicken, leg portion' the weight of the bone is included in the weight of 190g. Thus, descriptions of the food and portion should help users of the Tables to identify the appropriate food item and also the portion details if the given portion size is used to adapt nutrient values for a modified quantity.

The nutrients provided by the specified portions are abbreviated in the table headings. For clarification of this, reference to the key, opposite, should be made.

No	Food and portion description		Portion weight	Energy		Pro	Fat	Carb	Fib	Minerals		Vitamins				
										Ca	Fe	A	B$_1$	B$_2$	B$_3$	C
			g	kcal	kJ	g	g	g	g	mg	mg	µg	mg	mg	mg	mg

BATTERS – SAVOURY

No	Food and portion description		Weight g	kcal	kJ	Pro g	Fat g	Carb g	Fib g	Ca mg	Fe mg	A µg	B$_1$ mg	B$_2$ mg	B$_3$ mg	C mg
1	Crumpets/pikelets, toasted	2 crumpets	60	119	508	4.0	0.6	26.0	1.7	72	0.7	0	0.10	0.02	1.4	0
2	Pancakes, *made with skimmed milk*															
		2 pancakes	70	174	727	4.5	10.3	16.9	0.7	91	0.6	18	0.08	0.13	1.4	1
3	*made with whole milk*	2 pancakes	70	191	797	4.4	12.3	16.8	0.7	91	0.6	44	0.08	0.13	1.4	1
4	Yorkshire puddings, *made with*															
	skimmed milk	2 individual puddings	50	93	390	3.4	3.7	12.4	0.5	65	0.5	15	0.05	0.08	1.1	1
5	*made with whole milk*	2 individual puddings	50	104	437	3.3	5.0	12.4	0.5	65	0.5	33	0.05	0.08	1.1	1

BEVERAGES – NON-ALCOHOLIC

Measures for average cup and mug

No	Food and portion description		Weight g	kcal	kJ	Pro g	Fat g	Carb g	Fib g	Ca mg	Fe mg	A µg	B$_1$ mg	B$_2$ mg	B$_3$ mg	C mg
6	Bournvita, *powder*	2 heaped teaspoons	9	34	144	0.8	0.5	7.1	N	8	0.2	Tr	N	N	N	0
7	Bovril, *spread*	1 heaped teaspoon	5	9	37	1.9	0	0.1	0	2	0.7	0	0.46	0.37	4.3	0
8	Cocoa, *powder*	1 level teaspoon	3	9	39	0.6	0.7	0.3	N	4	0.3	0	0	0	0.2	0
9	Coffee, *ground/infused*	1 cup or mug	195	4	16	0.4	0	0.6	N	4	0	0	0	0.02	1.4	0
10	*instant, powder/granules*	1 heaped teaspoon	2	2	8	0.3	0	0.2	0.1	3	0.1	0	0	0	0.5	0
11	Drinking chocolate, *powder*	3 heaped teaspoons	15	55	233	0.8	0.9	11.6	N	5	0.4	N	0.01	0.01	0.3	0
12	Horlick's, *powder*	4 heaped teaspoons	20	76	321	2.5	0.8	15.6	N	86	0.3	125	0.20	0.27	3.6	0
13	Marmite, *spread*	1 heaped teaspoon	5	9	38	2.0	0	0.1	N	5	0.2	0	0.16	0.55	3.4	0
14	Milk shake, *powder*	3 heaped teaspoons	18	70	298	0.2	0.3	17.7	0	1	0.4	Tr	Tr	0	0.1	0
15	Ovaltine, *powder*	4 heaped teaspoons	15	54	228	1.4	0.4	11.9	N	12	0.3	94	0.15	0.20	2.6	0
16	Tea, *infused*	1 cup or mug	195	Tr	Tr	0.2	Tr	Tr	0	Tr	Tr	0	Tr	0.02	0.2	0

BISCUITS – PLAIN

No	Food and portion description		Portion weight (g)	Energy kcal	Energy kJ	Pro g	Fat g	Carb g	Fib g	Minerals Ca mg	Minerals Fe mg	Vitamins A µg	Vitamins B₁ mg	Vitamins B₂ mg	Vitamins B₃ mg	Vitamins C mg
17	Crackers, cream	3 crackers	21	92	390	2.0	3.4	14.3	1.3	23	0.4	0	0.05	0.01	0.8	0
18	wholemeal	3 crackers	21	87	366	2.1	2.4	15.1	1.0	23	0.5	0	0.05	0.01	1.0	0
19	Crispbread, cracotte type	3 crispbreads	15	61	256	1.3	2.3	9.4	1.4	12	0.3	0	0.03	0.01	0.5	0
20	rye	3 crispbreads	24	77	328	2.3	0.5	16.9	2.8	11	0.8	0	0.07	0.03	0.7	0
21	Matzos	1 matzo	30	115	490	3.2	0.6	26.0	1.1	10	0.5	0	0.03	0.01	0.9	0
22	Oatcakes	2 oatcakes	26	115	482	2.6	4.8	16.4	0.9	14	1.2	0	0.08	0.02	0.8	0
23	Water biscuits	3 biscuits	21	92	390	2.3	2.6	15.9	1.3	25	0.3	0	0.02	0.01	0.7	0

BISCUITS – SWEET

No	Food and portion description		Portion weight (g)	Energy kcal	Energy kJ	Pro g	Fat g	Carb g	Fib g	Minerals Ca mg	Minerals Fe mg	Vitamins A µg	Vitamins B₁ mg	Vitamins B₂ mg	Vitamins B₃ mg	Vitamins C mg
24	Chocolate biscuits, *e.g.* Club, Penguin	1 biscuit	25	131	549	1.4	6.9	16.9	0.7	28	0.4	Tr	0.01	0.03	0.4	0
25	Digestive biscuits, chocolate	2 biscuits	30	148	621	2.0	7.2	20.0	0.9	25	0.6	Tr	0.02	0.03	0.8	0
26	plain	2 biscuits	30	141	593	1.9	6.3	20.6	1.4	28	1.0	0	0.04	0.03	0.7	0
27	Ginger nuts	2 biscuits	20	91	385	1.1	3.0	15.8	0.4	26	0.8	N	0.02	0.01	0.4	0
28	Sandwich biscuits, *e.g.* Bourbon, custard creams	2 biscuits	25	128	538	1.3	6.5	17.3	0.3	25	0.4	0	0.04	0.03	0.5	0
29	Semi-sweet biscuits, *e.g.* Marie, rich tea	2 biscuits	15	69	289	1.0	2.5	11.2	0.3	18	0.3	0	0.02	0.01	0.4	0
30	Short-sweet biscuits, *e.g.* Lincoln, shortcake	2 biscuits	20	94	393	1.2	4.7	12.4	0.3	17	0.4	0	0.03	0.01	0.4	0
31	Shortbread	2 fingers	35	174	730	2.1	9.1	22.4	0.8	32	0.5	95	0.05	0.01	0.8	0
32	Wafers, filled	3 wafers	18	96	404	0.8	5.4	11.9	0.3	13	0.3	0	0.02	0.01	0.3	0

BREAD AND BREAD ROLLS – PLAIN

No	Food and portion description	Portion weight		Energy		Pro	Fat	Carb	Fib	Minerals			Vitamins			
										Ca	Fe	A	B₁	B₂	B₃	C
			g	kcal	kJ	g	g	g	g	mg	mg	µg	mg	mg	mg	mg
33	**Bread**, brown, from large loaf, medium sliced	2 slices	70	153	649	6.0	1.4	31.0	4.1	70	1.5	0	0.19	0.06	2.9	0
34	white, from large loaf, medium sliced	2 slices	75	176	752	6.3	1.4	37.0	2.9	83	1.2	0	0.16	0.05	2.6	0
35	wholemeal, from large loaf, medium sliced	2 slices	70	151	640	6.4	1.8	29.1	5.2	38	1.9	0	0.24	0.06	4.1	0
36	**Bread rolls**, brown bap	1 bap	55	147	626	5.5	2.1	28.5	3.5	61	1.9	0	0.23	0.04	3.0	0
37	white bap	1 bap	55	147	625	5.1	2.3	28.4	2.1	66	1.2	0	0.15	0.02	2.1	0
38	wholemeal bap	1 bap	55	133	564	5.0	1.6	26.6	4.8	30	1.9	0	0.17	0.05	3.2	0
39	**Chapatis**, made with fat	1 chapati	70	230	968	5.7	9.0	33.8	4.9	46	1.6	N	0.18	0.03	2.4	0
40	made without fat	1 chapati	70	141	602	5.1	0.7	30.6	4.5	42	1.5	0	0.16	0.03	2.1	0
41	**Naan**	1 naan	170	571	2406	15.1	21.3	85.2	3.7	272	2.2	165	0.32	0.17	5.1	Tr
42	**Papadums**, fried	2 papadums	22	81	341	3.9	3.7	8.6	2.0	15	2.4	N	0.03	0.02	0.7	0
43	**Paratha**	1 paratha	125	403	1695	10.0	17.9	54.0	5.5	105	2.5	175	0.23	0.05	5.1	0
44	**Pitta bread**, made with white flour	1 pitta	65	172	733	6.0	0.8	37.6	2.5	59	1.1	0	0.16	0.03	2.1	0
45	made with wholemeal flour	1 pitta	65	159	678	5.5	0.7	34.8	5.9	31	1.8	0	0.16	0.03	2.1	0
	BREAD AND BUNS – SWEET															
46	**Bath bun**	1 bun	55	201	848	4.3	7.6	30.9	1.2	61	0.8	9	0.09	0.07	1.7	Tr
47	**Chelsea bun**	1 bun	70	256	1079	5.5	9.7	39.3	1.5	77	1.1	12	0.11	0.09	2.1	Tr
48	**Currant bread**	2 slices	50	145	610	3.8	3.8	25.4	1.9	43	0.8	Tr	0.10	0.05	1.5	0
49	**Currant bun**	1 bun	50	148	625	3.8	3.8	26.4	0.9	55	1.0	N	0.19	0.08	1.6	0
50	**Hot cross bun**	1 bun	65	202	853	4.8	4.4	38.0	1.4	72	1.0	44	0.10	0.07	1.8	0
51	**Malt loaf**	2 slices	60	161	683	5.0	1.4	34.1	3.9	66	1.7	Tr	0.27	0.08	2.7	0

BREAKFAST CEREALS

No	Food and portion description		Portion weight (g)	Energy kcal	Energy kJ	Pro g	Fat g	Carb g	Fib g	Minerals Ca mg	Minerals Fe mg	Vitamins A µg	Vitamins B₁ mg	Vitamins B₂ mg	Vitamins B₃ mg	Vitamins C mg
52	All-Bran	1 serving	45	113	482	6.8	1.5	19.4	13.5	31	5.4	0	0.45	0.68	8.6	0
53	Bran	2 tablespoons	12	25	105	1.7	0.7	3.2	4.8	13	1.5	0	0.11	0.04	3.9	0
54	Bran Buds	1 serving	75	205	872	9.8	2.2	39.0	21.2	42	9.0	0	0.75	1.13	14.1	0
55	Bran Flakes	1 serving	45	144	612	4.6	0.9	31.4	7.8	23	18.0	0	0.45	0.68	8.3	16
56	Coco Pops	1 serving	35	135	575	1.9	0.5	32.9	0.4	12	2.3	0	0.35	0.53	6.0	0
57	Corn Flakes	1 serving	25	89	379	2.0	0.2	21.2	0.9	4	1.7	0	0.25	0.38	4.2	0
58	Crunchy Nut Corn Flakes	1 serving	45	179	761	3.3	1.8	39.9	0.7	8	3.0	0	0.45	0.68	7.6	N
59	Frosties	1 serving	45	172	734	2.4	0.2	42.9	0.5	5	3.0	0	0.45	0.68	7.5	0
60	Fruit 'n' Fibre	1 serving	50	176	748	4.1	2.6	36.6	5.1	26	3.4	Tr	0.50	0.75	8.9	N
61	Grapenuts	1 serving	90	311	1328	9.5	0.5	71.9	5.6	33	8.6	1188	1.17	1.35	18.2	0
62	Muesli	1 serving	95	346	1469	10.1	5.6	67.5	7.7	114	5.3	Tr	0.48	0.67	8.4	Tr
63	with extra fruit	1 serving	95	353	1498	9.7	5.9	69.7	5.1	63	3.8	N	0.10	0.48	6.8	2
64	with no added sugar	1 serving	95	348	1474	10.0	7.7	63.7	10.5	47	3.2	Tr	0.28	0.28	7.1	Tr
65	Porridge, made with milk	1 serving	160	186	781	7.7	8.2	21.9	1.3	192	1.0	90	0.16	0.27	2.1	2
66	made with milk and water	1 serving	160	133	557	5.1	5.0	18.1	1.3	106	0.8	45	0.13	0.14	1.4	Tr
67	made with water	1 serving	160	78	334	2.4	1.8	14.4	1.3	11	0.8	0	0.10	0.02	0.6	0
68	Puffed Wheat	1 serving	20	64	273	2.8	0.3	13.5	1.8	5	0.9	0	Tr	0.01	1.6	0
69	Ready Brek	1 serving	30	117	494	3.7	2.6	20.9	2.0	20	1.4	0	0.45	0.03	3.7	0
70	Rice Krispies	1 serving	35	129	550	2.1	0.3	31.4	0.4	7	2.3	0	0.35	0.53	6.1	0
71	Ricicles	1 serving	45	172	734	1.9	0.2	43.3	0.4	9	3.0	0	0.45	0.68	7.7	0

No	Food and portion description		Portion weight g	Energy kcal	Energy kJ	Pro g	Fat g	Carb g	Fib g	Minerals Ca mg	Minerals Fe mg	Vitamins A µg	Vitamins B₁ mg	Vitamins B₂ mg	Vitamins B₃ mg	Vitamins C mg

BREAKFAST CEREALS

No	Food and portion description		Portion weight g	kcal	kJ	Pro g	Fat g	Carb g	Fib g	Ca mg	Fe mg	A µg	B₁ mg	B₂ mg	B₃ mg	C mg
72	Shedded Wheat	2 pieces	45	146	623	4.8	1.4	30.7	4.5	17	1.9	0	0.12	0.02	3.0	0
73	Shreddies	1 serving	55	182	776	5.5	0.8	40.8	6.0	22	1.5	0	0.66	1.21	12.7	0
74	Special K	1 serving	35	133	566	5.4	0.4	28.9	0.9	25	4.7	0	0.42	0.60	7.4	0
75	Sugar Puffs	1 serving	50	162	691	3.0	0.4	42.3	2.4	7	1.1	0	Tr	0.02	1.9	0
76	Sultana Bran	1 serving	35	106	450	3.0	0.6	23.7	5.4	18	10.5	Tr	0.35	0.53	6.3	N
77	Weetabix	2 Weetabix	40	142	604	4.3	0.8	31.3	3.4	14	2.4	0	0.28	0.40	4.8	0
78	Weetaflakes	1 serving	45	154	654	4.8	0.9	33.7	5.2	16	2.7	0	0.09	0.45	5.4	0
79	Wheatgerm	2 heaped tablespoons	15	45	191	4.0	1.4	6.7	2.3	8	1.3	0	0.30	0.11	1.5	0

CAKES AND PASTRIES

No	Food and portion description		Portion weight g	kcal	kJ	Pro g	Fat g	Carb g	Fib g	Ca mg	Fe mg	A µg	B₁ mg	B₂ mg	B₃ mg	C mg
80	All-Bran loaf	1 slice	80	203	865	4.2	1.3	46.7	4.3	68	2.0	13	0.14	0.19	2.9	0
81	Baclava	1 baclava	110	354	1484	5.2	18.7	44.0	2.1	48	1.0	N	0.10	0.04	2.2	N
82	Battenburg cake	1 slice	55	204	853	3.2	9.6	27.5	0.8	48	0.6	25	0.04	0.09	1.0	0
83	Cheesecake, *frozen*	1 slice	100	242	1017	5.7	10.6	33.0	0.9	68	0.5	N	0.04	0.16	1.7	0
84	Chocolate cake, *with butter icing*	1 slice	40	192	804	2.3	11.9	20.4	1.2	23	0.6	120	0.03	0.04	0.8	0
85	Chocolate éclair	1 éclair	40	149	624	1.6	9.5	15.2	0.2	19	0.4	84	0.02	0.04	0.5	Tr
86	Cream horn	1 cream horn	60	261	1082	2.3	21.5	15.5	0.6	37	0.4	126	0.04	0.04	0.8	1
87	Crispie cake	2 crispie cakes	30	139	585	1.7	5.6	21.9	0.5	9	1.2	1	0.12	0.18	2.1	0
88	Croissant	1 croissant	50	180	753	4.2	10.2	19.2	1.3	40	1.0	11	0.09	0.08	1.9	0
89	Custard tart	1 individual tart	80	222	929	5.0	11.6	25.9	1.0	76	0.6	26	0.11	0.13	1.5	0

CAKES AND PASTRIES

No	Food and portion description		Portion weight (g)	Energy kcal	Energy kJ	Pro g	Fat g	Carb g	Fib g	Minerals Ca mg	Minerals Fe mg	Vitamins A µg	B₁ mg	B₂ mg	B₃ mg	C mg
90	**Danish pastry**	1 Danish pastry	**100**	374	1571	5.8	17.6	51.3	2.7	92	1.3	N	0.13	0.07	2.1	0
91	**Doughnuts**, *jam filled*	1 doughnut	**70**	235	990	4.0	10.2	34.2	1.8	50	0.8	N	0.15	0.05	1.8	N
92	ring	1 doughnut	**50**	199	831	3.1	10.9	23.6	1.6	38	0.6	N	0.11	0.04	1.2	0
93	ring, iced	1 doughnut	**60**	230	966	2.9	10.5	33.1	1.4	35	0.6	N	0.10	0.04	1.5	0
94	**Eccles cake**	1 Eccles cake	**60**	285	1195	2.3	15.8	35.6	1.2	47	0.7	37	0.07	0.02	1.0	0
95	**Flapjack**	1 flapjack	**30**	145	608	1.4	8.0	18.1	0.8	11	0.6	69	0.08	0.01	0.4	0
96	**Fondant fancy**	1 fondant fancy	**25**	102	429	1.0	3.7	17.2	0.6	11	0.4	N	0	0.01	0.3	0
97	**Fruit cake**, *made with white flour*	1 slice	**60**	212	894	3.1	7.7	34.7	1.5	36	1.0	N	0.05	0.04	1.0	0
98	*made with wholemeal flour*	1 slice	**60**	218	915	3.6	9.4	31.7	1.8	51	1.1	96	0.07	0.05	1.6	0
99	**Fruit pie**	1 individual pie	**110**	406	1709	4.7	17.1	62.4	2.5	56	1.3	Tr	0.06	0.02	1.4	Tr
100	**Gateau**	1 slice	**45**	152	636	2.6	7.6	19.5	0.2	27	0.4	117	0.03	0.08	0.8	0
101	**Gingerbread**	1 slice	**65**	246	1038	3.7	8.2	42.1	0.9	53	1.0	85	0.07	0.05	1.2	0
102	**Jaffa cake**	2 Jaffa cakes	**20**	73	306	0.7	2.1	13.6	N	11	0.3	3	0.01	0.01	0.2	0
103	**Jam tarts**, *made with white flour pastry*	1 jam tart	**35**	129	543	1.2	4.6	22.2	0.9	25	0.6	N	0.02	0.01	0.4	Tr
104	*made with wholemeal flour pastry*	1 jam tart	**35**	129	544	1.5	5.3	20.1	1.2	8	0.7	22	0.04	0.01	0.9	1
105	**Madeira cake**	1 slice	**25**	98	413	1.4	4.2	14.6	0.3	11	0.3	N	0.02	0.03	0.4	0
106	**Meringue**, *filled with dairy cream*	1 meringue	**35**	132	550	1.2	8.3	14.0	0	14	Tr	128	0	0.07	0.3	Tr
107	**Mince pie**	1 individual pie	**50**	212	886	2.2	10.2	29.5	1.4	38	0.8	41	0.06	0.01	0.8	0
108	**Muffins**, bran	1 muffin	**70**	190	804	5.5	5.4	31.9	6.0	84	2.3	57	0.15	0.11	5.3	Tr
109	plain	1 muffin	**70**	198	839	7.1	4.4	34.7	1.9	98	1.3	47	0.14	0.11	2.7	Tr

CAKES AND PASTRIES

No	Food and portion description		Portion weight	Energy		Pro	Fat	Carb	Fib	Minerals		Vitamins				
										Ca	Fe	A	B₁	B₂	B₃	C
			g	kcal	kJ	g	g	g	g	mg	mg	μg	mg	mg	mg	mg
110	**Rock cake**	1 rock cake	80	317	1332	4.3	13.1	48.4	1.4	88	1.0	136	0.10	0.06	1.4	0
111	**Scotch pancake**	1 pancake	30	88	368	1.7	3.5	13.1	0.5	36	0.3	36	0.04	0.03	0.6	Tr
112	**Sponge cake**, *with jam filling*	1 slice	35	106	448	1.5	1.7	22.5	0.4	15	0.6	N	0.01	0.02	0.5	0
113	*with butter icing filling*	1 slice	35	172	716	1.6	10.7	18.3	0.2	16	0.3	112	0.02	0.03	0.5	0
114	**Swiss roll**	1 slice	35	97	410	2.5	1.5	19.4	0.4	34	0.5	28	0.02	0.06	0.8	0
115	*chocolate mini roll*	1 mini roll	25	84	355	1.1	2.8	14.5	0.6	19	0.3	N	0.03	0.05	0.3	0
116	**Vanilla slice**	1 vanilla slice	75	248	1037	3.4	13.4	30.2	0.7	59	0.6	12	0.06	0.07	1.1	1

CHEESES

No	Food and portion description		Portion weight	Energy		Pro	Fat	Carb	Fib	Minerals		Vitamins				
			g	kcal	kJ	g	g	g	g	mg	mg	μg	mg	mg	mg	mg
117	**Brie**	1 slice	40	128	529	7.7	10.8	Tr	0	216	0.3	128	0.02	0.17	2.0	Tr
118	**Caerphilly**	1 slice	40	150	622	9.3	12.5	0	0	220	0.3	140	0.01	0.19	2.2	Tr
119	**Camembert**	1 slice	40	119	493	8.4	9.5	Tr	0	140	0.1	113	0.02	0.21	2.3	Tr
120	**Cheddar**	1 slice	40	165	683	10.2	13.8	0	0	288	0.1	145	0.01	0.16	2.4	Tr
121	**Cheddar-type**, *reduced fat*	1 slice	40	104	436	12.6	6.0	Tr	0	336	0.1	73	0.01	0.21	3.0	Tr
122	**Cheshire**	1 slice	40	152	628	9.6	12.6	0	0	224	0.1	155	0.01	0.19	2.3	Tr
123	**Cheshire-type**, *reduced fat*	1 slice	40	108	449	13.1	6.1	Tr	0	260	0.1	65	0.02	0.22	3.1	Tr
124	**Cottage cheese**	1 serving	45	44	186	6.2	1.8	0.9	0	33	0	21	0.01	0.12	1.5	Tr
125	*reduced fat*	1 serving	45	35	149	6.0	0.6	1.5	0	33	0	8	0.01	0.12	1.5	Tr
126	**Cream cheese**	1 serving	30	132	542	0.9	14.2	Tr	0	29	0	127	0.01	0.04	0.2	Tr
127	**Danish blue**	1 slice	40	139	575	8.0	11.8	Tr	0	200	0.1	129	0.01	0.16	2.1	Tr
128	**Derby**	1 slice	40	161	667	9.7	13.6	0	0	272	0.2	151	0.01	0.16	2.3	Tr
129	**Double Gloucester**	1 slice	40	162	671	9.8	13.6	0	0	264	0.2	151	0.01	0.18	2.4	Tr

CHEESES

No	Food and portion description	Portion weight		Energy		Pro	Fat	Carb	Fib	Minerals		Vitamins				
										Ca	Fe	A	B$_1$	B$_2$	B$_3$	C
			g	kcal	kJ	g	g	g	g	mg	mg	µg	mg	mg	mg	mg
130	Edam	1 slice	40	133	553	10.4	10.2	Tr	0	308	0.2	80	0.01	0.14	2.5	Tr
131	Edam-type, reduced fat	1 slice	40	92	383	13.0	4.4	Tr	0	N	N	34	N	N	N	Tr
132	Emmental	1 slice	40	153	635	11.5	11.9	Tr	0	388	0.1	137	0.02	0.14	2.7	Tr
133	Feta	1 slice	40	100	415	6.2	8.1	0.6	0	144	0.1	90	0.02	0.08	1.5	Tr
134	Fromage frais	1 serving	45	51	211	3.1	3.2	2.6	0	40	0	45	0.02	0.18	0.8	Tr
135	Gouda	1 slice	40	150	622	9.6	12.4	Tr	0	296	0	108	0.01	0.12	2.3	Tr
136	Gruyere	1 slice	40	164	678	10.9	13.3	Tr	0	380	0.1	145	0.01	0.16	2.6	Tr
137	Lancashire	1 slice	40	149	618	9.3	12.4	0	0	224	0.1	144	0.01	0.18	2.2	Tr
138	Leicester	1 slice	40	160	665	9.7	13.5	0	0	264	0.2	146	0.01	0.18	2.3	Tr
139	Lymeswold	1 slice	40	170	702	6.2	16.1	Tr	0	108	0.1	198	0.02	0.17	1.7	Tr
140	Parmesan	2 heaped teapoons	9	41	169	3.5	2.9	Tr	0	108	0.1	34	0	0.04	0.8	Tr
141	Processed cheese	1 slice	20	66	273	4.2	5.4	0.2	0	120	0.1	57	0.01	0.06	1.0	Tr
142	Ricotta	1 serving	30	43	180	2.8	3.3	0.6	0	72	0.1	60	0.01	0.06	0.7	Tr
143	Roquefort	1 slice	40	150	621	7.9	13.2	Tr	0	212	0.2	119	0.02	0.26	2.1	Tr
144	Sage Derby	1 slice	40	161	667	9.7	13.6	0	0	244	0.3	152	0.01	0.17	2.3	Tr
145	Soya cheese	1 slice	40	128	528	7.3	10.9	Tr	0	180	0.4	0	0.10	0.25	1.6	0
146	Stilton, blue	1 slice	40	164	680	9.1	14.2	0	0	128	0.1	154	0.01	0.17	2.3	Tr
147	white	1 slice	40	145	599	8.0	12.5	0	0	100	0.1	137	0.01	0.15	1.9	Tr
148	Wensleydale	1 slice	40	151	625	9.3	12.6	0	0	224	0.1	127	0.01	0.18	2.2	Tr

CHEESE DISHES

No	Food and portion description		Portion weight	Energy		Pro	Fat	Carb	Fib	Minerals		Vitamins				
										Ca	Fe	A	B₁	B₂	B₃	C
			g	kcal	kJ	g	g	g	g	mg	mg	µg	mg	mg	mg	mg
149	**Cauliflower cheese**	1 serving	**310**	326	1358	18.3	21.4	15.8	4.3	372	1.9	236	0.31	0.31	5.9	22
150	**Cheese and potato pie**	1 serving	**155**	215	905	7.4	12.6	19.5	0.9	136	0.6	147	0.14	0.14	2.3	5
151	**Cheese omelette**	2 eggs	**195**	519	2157	31.0	44.1	Tr	0	546	2.3	550	0.12	0.62	8.4	Tr
152	**Cheese pudding**	1 serving	**115**	196	815	11.6	12.5	9.8	0.5	253	0.8	159	0.06	0.28	3.2	Tr
153	**Cheese soufflé**	1 serving	**100**	253	1053	11.4	19.2	9.3	0.3	210	1.0	243	0.07	0.26	3.1	Tr
154	**Macaroni cheese**	1 serving	**180**	320	1337	13.1	19.4	24.5	1.4	306	0.7	220	0.07	0.29	3.6	Tr
155	**Pizza**	1 slice	**160**	376	1574	14.4	18.9	39.7	2.9	304	1.6	122	0.16	0.21	4.6	5
156	**Quiche, cheese,** *made with white flour pastry*	1 slice	**90**	283	1179	11.3	20.0	15.6	0.6	234	0.9	182	0.07	0.21	3.2	Tr
157	*made with wholemeal flour pastry*	1 slice	**90**	277	1155	11.9	20.2	13.1	1.6	216	1.3	182	0.09	0.22	4.0	Tr
158	**Lorraine,** *made with white flour pastry*	1 slice	**90**	352	1466	14.5	25.3	17.8	0.7	207	1.1	158	0.14	0.23	4.8	Tr
159	*made with wholemeal flour pastry*	1 slice	**90**	346	1439	15.2	25.5	14.9	1.8	180	1.4	158	0.16	0.23	5.6	Tr
160	**mushroom,** *made with white flour pastry*	1 slice	**90**	256	1067	9.0	17.6	16.5	1.1	180	0.9	149	0.08	0.20	3.2	Tr
161	*made with wholemeal flour pastry*	1 slice	**90**	249	1040	9.7	17.7	13.7	2.1	153	1.3	149	0.10	0.21	4.0	Tr
162	**Welsh rarebit,** *made with white bread*	1 slice toast	**60**	228	952	8.3	15.8	14.3	1.1	204	0.5	165	0.05	0.10	2.5	Tr
163	*made with wholemeal bread*	1 slice toast	**60**	221	925	8.6	15.4	13.3	2.3	180	0.9	155	0.08	0.10	3.1	Tr

No	Food and portion description		Portion weight	Energy		Pro	Fat	Carb	Fib	Minerals		Vitamins				
										Ca	Fe	A	B_1	B_2	B_3	C
			g	kcal	kJ	g	g	g	g	mg	mg	µg	mg	mg	mg	mg
	CHOCOLATE															
164	**Bounty**	2 small bars	**60**	284	1188	2.9	15.7	35.0	N	66	0.8	4	0.02	0.06	0.7	0
165	**Chocolate**, fancy/filled	1 bar	**55**	253	1066	2.3	10.3	40.3	N	51	1.0	4	0.06	0.06	0.6	0
166	milk	1 bar	**50**	265	1107	4.2	15.2	29.7	N	110	0.8	4	0.05	0.12	0.8	0
167	plain	1 bar	**50**	263	1099	2.4	14.6	32.4	N	19	1.2	4	0.04	0.04	0.6	0
168	**Mars**	1 bar	**65**	287	1204	3.4	12.3	43.2	N	104	0.7	5	0.03	0.13	0.8	0
	CHUTNEY AND PICKLES															
169	**Apple chutney**	1 serving	**35**	68	288	0.2	0	17.7	0.6	9	0.3	1	0.01	0.01	0.1	1
170	**Piccalilli**	1 serving	**40**	13	56	0.4	0.3	2.4	0.7	10	0.4	N	0.06	0	0.2	0
171	**Sweet pickle**, *e.g. Branston, Pan Yan*	1 serving	**35**	47	200	0.2	0.1	12.0	0.5	7	0.7	N	0.01	0	0.1	N
172	**Tomato chutney**	1 serving	**35**	54	230	0.4	0	13.9	0.6	11	0.4	21	0.01	0.02	0.2	3
	COBBLERS AND DUMPLINGS															
173	**Cobblers**, *made with white flour*	2 cobblers	**50**	181	762	3.6	7.3	26.9	1.1	90	0.7	70	0.08	0.04	1.3	Tr
174	*made with wholemeal flour*	2 cobblers	**50**	163	684	4.4	7.2	21.6	2.5	55	1.2	65	0.11	0.05	2.5	Tr
175	**Dumplings**	2 dumplings	**100**	208	871	2.8	11.7	24.5	1.0	52	0.6	9	0.05	0.01	0.9	0

No	Food and portion description	Portion weight	Energy		Pro	Fat	Carb	Fib	Minerals		Vitamins				
									Ca	Fe	A	B$_1$	B$_2$	B$_3$	C
		g	kcal	kJ	g	g	g	g	mg	mg	µg	mg	mg	mg	mg
CREAM – IN DRINKS AND SOUPS															
	Average in cup, mug and soup bowl														
176	**Double cream**	1 serving **25**	112	462	0.4	12.0	0.7	0	13	0.1	164	0.01	0.04	0.1	0
177	**Single cream**	1 serving **25**	50	204	0.7	4.8	1.0	0	23	0	84	0.01	0.04	0.2	0
178	**Whipping cream**	1 serving **25**	93	385	0.5	9.8	0.8	0	16	Tr	152	0.01	0.04	0.1	0
CREAM – ON PUDDINGS															
	On a medium portion														
179	**Double cream**	1 serving **35**	157	647	0.6	16.8	0.9	0	18	0.1	229	0.01	0.06	0.2	0
180	**Single cream**	1 serving **35**	69	286	0.9	6.7	1.4	0	32	0	118	0.01	0.06	0.2	0
181	**Whipping cream**	1 serving **35**	131	539	0.7	13.8	1.1	0	22	Tr	213	0.01	0.06	0.2	0
EGGS															
182	**Boiled egg**	1 size 2 **60**	88	367	7.5	6.5	Tr	0	34	1.1	114	0.04	0.21	2.3	0
183	**Fried egg**	1 size 2 **60**	107	447	8.2	8.3	Tr	0	39	1.3	129	0.04	0.19	2.4	0
184	**Poached egg**	1 size 2 **60**	88	367	7.5	6.5	Tr	0	34	1.1	114	0.04	0.22	2.3	0
EGG DISHES															
185	**Egg and bacon pie**	1 slice **155**	463	1934	19.8	29.8	31.2	1.9	105	2.6	171	0.29	0.37	7.1	0
186	**Egg fried rice**	1 serving **190**	395	1659	8.0	20.1	48.8	1.7	25	1.0	67	0.06	0.15	2.7	Tr
187	**Omelette**	2 eggs **135**	258	1069	14.7	22.1	Tr	0	69	2.3	325	0.09	0.45	4.5	0
188	**Scotch egg**	1 egg **120**	301	1255	14.4	20.5	15.7	1.9	60	2.2	36	0.10	0.25	4.7	N
189	**Scrambled egg**	2 eggs **140**	346	1435	15.0	31.6	0.8	0	88	2.2	430	0.10	0.46	4.5	Tr
190	**Soufflé**, plain	1 serving **100**	201	838	7.6	14.7	10.4	0.4	95	1.0	194	0.07	0.22	2.3	Tr

No	Food and portion description		Portion weight	Energy		Pro	Fat	Carb	Fib	Minerals		Vitamins				
										Ca	Fe	A	B₁	B₂	B₃	C
			g	kcal	kJ	g	g	g	g	mg	mg	µg	mg	mg	mg	mg

FAT – ON BREAD

Average on 1 slice bread/large loaf and both sides bread roll

No	Food and portion description		Portion weight g	kcal	kJ	Pro g	Fat g	Carb g	Fib g	Ca mg	Fe mg	A µg	B₁ mg	B₂ mg	B₃ mg	C mg
191	**Butter**	Medium layer	8	59	242	0	6.5	Tr	0	1	0	71	Tr	0	0	Tr
192	**Dairy/fat spread**	Medium layer	8	53	218	0	5.9	Tr	0	1	Tr	75	Tr	Tr	0	0
193	**Low fat spread**	Medium layer	8	31	128	0.5	3.2	0	0	3	Tr	87	Tr	Tr	0.1	0
194	**Margarine**	Medium layer	8	59	243	0	6.5	0.1	0	0	0	72	Tr	Tr	Tr	0
195	**Very low fat spread**	Medium layer	8	22	90	0.7	2.0	0.3	0	N	N	N	Tr	Tr	0.2	0

FAT – ON CRACKERS, CRISPBREADS, ETC.

Average on 1 biscuit

196	**Butter**	Medium layer	3	22	91	0	2.5	Tr	0	0	0	27	Tr	0	0	Tr
197	**Dairy/fat spread**	Medium layer	3	20	82	0	2.2	Tr	0	0	Tr	28	Tr	Tr	0	0
198	**Low fat spread**	Medium layer	3	12	48	0.2	1.2	0	0	1	Tr	33	Tr	Tr	0	0
199	**Margarine**	Medium layer	3	22	91	0	2.4	0	0	0	0	27	Tr	Tr	Tr	0
200	**Very low fat spread**	Medium layer	3	8	34	0.2	0.8	0.1	0	N	N	N	Tr	Tr	0.1	0

FAT – WITH JACKET POTATO

On medium size jacket potato

201	**Butter**	1 chunk	10	74	303	0.1	8.2	Tr	0	2	0	89	Tr	0	0	Tr
202	**Dairy/fat spread**	1 chunk	10	66	272	0	7.3	Tr	0	1	Tr	94	Tr	Tr	0	0
203	**Low fat spread**	1 chunk	10	39	161	0.6	4.1	0.1	0	4	Tr	108	Tr	Tr	0.1	0
204	**Margarine**	1 chunk	10	74	304	0	8.2	0.1	0	0	0	91	Tr	Tr	Tr	0
205	**Very low fat spread**	1 chunk	10	27	113	0.8	2.5	0.4	0	N	N	N	Tr	Tr	0.2	0

FISH

No	Food and portion description	Portion weight (g)	Energy kcal	Energy kJ	Pro g	Fat g	Carb g	Fib g	Minerals Ca mg	Fe mg	A µg	Vitamins B₁ mg	B₂ mg	B₃ mg	C mg
206	**Cod**, *in batter, fried*	1 piece **85**	169	709	16.7	8.8	6.4	0.3	68	0.4	0	0.06	0.06	4.6	0
207	*roe, in crumbs, fried* Chip shop portion	**80**	162	675	16.7	9.5	2.4	0.1	14	1.3	120	1.04	0.72	4.2	21
208	*steaks, grilled* 2 steaks	**130**	124	523	27.0	1.7	0	0	13	0.5	0	0.10	0.08	7.5	0
209	**Crab**, *white meat, canned* 1 serving	**70**	57	239	12.7	0.6	0	0	84	2.0	0	0	0.04	3.2	0
210	**Herring**, *fillets in oatmeal, fried* 2 fillets	**110**	257	1073	25.4	16.6	1.7	1.4	43	2.1	54	0	0.20	9.1	0
211	**Kipper**, *fillets, baked* 2 fillets	**130**	267	1112	33.2	14.8	0	0	85	1.8	64	0	0.23	11.4	0
212	**Mackerel**, *fillets, fried* 2 fillets	**110**	207	862	23.7	12.4	0	0	31	1.3	57	0.10	0.42	14.0	0
213	**Pilchards**, *in tomato sauce, canned* 1 serving	**105**	132	558	19.7	5.7	0.7	Tr	315	2.8	0	0.02	0.30	11.7	0
214	**Plaice**, *fillets, steamed* 2 small fillets	**120**	112	470	22.7	2.3	0	0	46	0.7	0	0.36	0.13	8.0	0
215	*fillet, in crumbs, fried* 1 fillet	**105**	239	999	18.9	14.4	9.0	0.4	70	0.8	0	0.24	0.19	6.6	0
216	**Prawns**, *peeled* 1 serving	**80**	86	361	18.1	1.4	0	0	120	0.9	Tr	N	N	N	Tr
217	**Rock salmon**, *in batter, fried* Chip shop portion	**205**	543	2261	34.2	38.5	15.8	0.6	86	2.3	N	0.12	0.21	17.8	0
218	**Salmon**, *cutlet, steamed* 1 cutlet	**135**	216	899	22.0	14.2	0	0	31	0.8	0	0.22	0.12	11.7	0
219	*red, canned, skin and bone removed* 1 serving	**115**	178	746	23.3	9.4	0	0	107	1.6	104	0.05	0.21	12.4	0
220	*smoked* 1 serving	**60**	85	359	15.2	2.7	0	0	11	0.4	0	0.10	0.10	8.1	0

No	Food and portion description	Portion weight (g)	Energy kcal	Energy kJ	Pro g	Fat g	Carb g	Fib g	Minerals Ca mg	Fe mg	Vitamins A µg	B1 mg	B2 mg	B3 mg	C mg
	FISH														
221	**Sardines**, *canned in oil, drained*	1 serving **70**	152	634	16.6	9.5	0	0	385	2.0	Tr	0.03	0.25	8.8	Tr
222	*canned in tomato sauce*	1 serving **85**	150	629	15.1	9.9	0.4	Tr	391	3.9	29	0.02	0.24	7.5	Tr
223	**Scampi**, *fried*	9 pieces **80**	253	1057	9.8	14.1	23.1	0.9	79	0.9	0	0.06	0.04	2.9	0
224	**Tuna**, *canned in oil, drained*	1 serving **95**	275	1142	21.7	20.9	0	0	7	1.0	N	0.04	0.10	16.3	0
	FISH DISHES AND PRODUCTS														
225	**Fish cakes**, fried	2 fish cakes **110**	207	864	10.0	11.6	16.6	0.7	77	1.1	0	0.07	0.07	3.1	0
226	**Fish curry**, *made with haddock*	1 serving **175**	445	1836	18.9	38.5	5.6	N	60	1.4	422	0.12	0.14	7.0	14
227	*made with herring*	1 serving **175**	611	2522	18.9	57.1	5.6	N	75	1.8	469	0.05	0.23	7.0	14
228	**Fish fingers**, fried	4 fish fingers **100**	233	975	13.5	12.7	17.2	0.6	45	0.7	0	0.08	0.07	3.9	0
229	**Fish pie**	1 serving **265**	339	1431	18.8	15.1	34.5	1.3	106	1.1	82	0.19	0.24	6.4	5
230	**Kedgeree**	1 serving **160**	242	1013	21.0	11.4	14.7	0.5	58	1.4	128	0.11	0.22	5.8	0
231	**Taramasalata**	1 serving **100**	446	1837	3.2	46.4	4.1	N	21	0.4	N	0.08	0.10	0	1
	FRUIT														
232	**Apple**	1 apple **120**	42	181	0.2	0	11.0	1.7	4	0.2	5	0.04	0.02	0.1	2
233	**Apricots**	3 apricots **110**	28	119	0.6	0	6.8	1.9	18	0.3	253	0.04	0.06	0.6	7
234	**Avocado pear**	½ pear **130**	201	830	3.8	20.0	1.6	1.6	14	1.4	15	0.09	0.09	1.6	14
235	**Banana**	1 banana **135**	63	273	0.9	0.3	15.4	2.4	5	0.3	27	0.03	0.05	0.7	8
236	**Blackberries**	15 blackberries **80**	23	100	1.0	0	5.1	5.3	50	0.7	14	0.02	0.03	0.5	16
237	**Cherries**	12 cherries **100**	41	175	0.5	0	10.4	1.4	14	0.3	17	0.04	0.06	0.4	4

FRUIT

No	Food and portion description	Portion weight (g)	Energy kcal	Energy kJ	Pro g	Fat g	Carb g	Fib g	Minerals Ca mg	Fe mg	A µg	Vitamins B₁ mg	B₂ mg	B₃ mg	C mg
238	**Fig**, green	1 fig **85**	35	148	1.1	Tr	8.1	2.0	29	0.3	71	0.05	0.04	0.5	2
239	**Gooseberries**	11 gooseberries **70**	26	110	0.4	0	6.4	2.2	13	0.4	21	0.03	0.02	0.3	28
240	**Grapes**, black	1 serving **140**	71	304	0.7	0	18.2	0.4	6	0.4	0	0.04	0.03	0.3	4
241	white	1 serving **140**	84	357	0.8	0	21.4	1.1	25	0.4	0	0.06	0.03	0.4	6
242	**Grapefruit**	½ grapefruit **140**	15	63	0.4	0	3.5	0.4	11	0.1	0	0.03	0.01	0.1	27
243	**Lemon**	1 wedge **25**	4	16	0.2	0	0.8	1.2	28	0.1	0	0.01	0.01	0.1	20
244	**Mango**	1 mango **315**	109	468	0.9	Tr	28.3	2.6	19	0.9	370	0.06	0.07	0.7	56
245	**Melon**, cantaloupe	½ melon **360**	54	227	2.2	0	11.9	1.8	43	1.8	709	0.11	0.07	1.1	54
246	honeydew	1 slice **190**	25	106	0.8	0	5.9	1.0	17	0.4	19	0.06	0.04	0.6	29
247	water	1 slice **320**	35	150	0.6	0	8.6	1.6	10	0.6	6	0.03	0.03	0.3	10
248	**Nectarine**	1 nectarine **110**	51	218	1.0	0	12.5	2.2	4	0.4	85	0.02	0.06	1.1	8
249	**Orange**	1 orange **245**	64	277	1.5	0	15.7	3.4	76	0.7	15	0.20	0.05	0.7	93
250	**Passion fruit**	4 passion fruit **170**	24	102	1.9	0	4.4	10.2	12	0.9	2	0	0.07	1.4	14
251	**Peach**	1 peach **125**	40	171	0.8	0	9.9	1.4	5	0.4	91	0.03	0.05	1.1	9
252	**Pear**	1 pear **150**	44	188	0.3	0	11.4	2.3	9	0.2	2	0.03	0.03	0.2	3
253	**Pineapple**	1 slice **125**	58	243	0.6	Tr	14.5	1.4	15	0.5	13	0.10	0.03	0.4	31

No	Food and portion description		Portion weight	Energy		Pro	Fat	Carb	Fib	Minerals		Vitamins				
										Ca	Fe	A	B_1	B_2	B_3	C
			g	kcal	kJ	g	g	g	g	mg	mg	µg	mg	mg	mg	mg
FRUIT																
254	**Plums**, dessert	3 plums	**105**	38	161	0.5	0	9.5	1.9	11	0.3	37	0.05	0.03	0.6	3
255	**Raspberries**	15 raspberries	**70**	18	74	0.6	0	3.9	4.7	29	0.8	9	0.01	0.02	0.4	18
256	**Strawberries**	1 serving	**100**	26	109	0.6	0	6.2	2.0	22	0.7	5	0.02	0.03	0.5	60
257	**Tangerine**	1 tangerine	**100**	23	100	0.6	0	5.6	1.2	29	0.2	12	0.05	0.01	0.2	21
FRUIT – CANNED IN SYRUP																
258	**Apricots**	5 halves	**140**	148	633	0.7	0	38.8	1.7	17	1.0	234	0.03	0.01	0.6	3
259	**Fruit salad**	1 serving	**130**	124	527	0.4	0	32.5	1.3	10	1.3	65	0.03	0.01	0.4	4
260	**Grapefruit**	6 segments	**120**	72	308	0.6	0	18.6	0.5	20	0.8	0	0.05	0.01	0.4	36
261	**Guavas**	6 halves	**175**	105	452	0.7	Tr	27.5	5.6	14	0.9	30	0.07	0.05	1.8	315
262	**Lychees**	10 lychees	**150**	102	435	0.6	Tr	26.6	0.6	6	1.1	Tr	0.05	0.05	0.5	12
263	**Mandarin oranges**	16 segments	**115**	64	273	0.7	Tr	16.3	0.3	21	0.5	9	0.08	0.02	0.3	16
264	**Mango**	2 slices	**135**	104	446	0.4	Tr	27.4	1.2	14	0.5	270	0.03	0.04	0.3	14
265	**Peaches**	6 slices	**110**	96	410	0.4	0	25.2	1.0	4	0.4	46	0.01	0.02	0.7	4
266	**Pears**	3 quarters	**135**	104	441	0.5	0	27.0	2.0	7	0.4	3	0.01	0.01	0.4	1
267	**Pineapple**	11 cubes	**150**	116	492	0.5	Tr	30.3	1.2	20	0.6	11	0.08	0.03	0.3	18
268	**Raspberries**	15 raspberries	**90**	78	333	0.5	0	20.3	4.1	13	1.5	12	0.01	0.03	0.4	6
269	**Strawberries**	10 strawberries	**85**	69	292	0.3	0	17.9	0.8	12	0.8	0	0.01	0.02	0.3	18

No	Food and portion description	Portion weight		Energy		Pro	Fat	Carb	Fib	Minerals		Vitamins				
										Ca	Fe	A	B₁	B₂	B₃	C
			g	kcal	kJ	g	g	g	g	mg	mg	µg	mg	mg	mg	mg
FRUIT – DRIED																
270	**Apricots**	8 apricots	**50**	91	388	2.4	0	21.7	10.8	46	2.1	300	0	0.10	1.9	0
271	**Currants**	2 handfuls	**35**	85	364	0.6	0	22.1	2.1	33	0.6	2	0.01	0.03	0.2	0
272	**Dates**	9 dates	**40**	99	422	0.8	Tr	25.6	3.1	27	0.6	3	0.03	0.02	1.2	0
273	**Figs**	4 figs	**60**	128	545	2.2	Tr	31.7	10.0	168	2.5	5	0.06	0.05	1.3	0
274	**Prunes**	8 prunes	**40**	54	228	0.8	0	13.4	4.8	12	1.0	55	0.03	0.07	0.6	0
275	**Raisins**	2 handfuls	**35**	86	367	0.4	0	22.5	2.1	21	0.6	2	0.04	0.03	0.2	0
276	**Sultanas**	2 handfuls	**35**	88	373	0.6	0	22.6	2.2	18	0.6	2	0.04	0.03	0.2	0
MEAT[a]																
277	**Bacon joints**, collar, boiled, *lean*	1 serving	**85**	162	681	22.1	8.2	0	0	13	1.6	0	0.31	0.26	7.2	0
278	collar, *boiled, lean and fat*	1 serving	**85**	276	1144	17.3	23.0	0	0	11	1.4	0	0.23	0.19	5.4	0
279	gammon, *boiled, lean*	1 serving	**85**	142	598	25.0	4.7	0	0	9	1.3	0	0.47	0.16	8.2	0
280	gammon, *boiled, lean and fat*	1 serving	**85**	229	951	21.0	16.1	0	0	8	1.1	0	0.37	0.13	6.8	0
281	**Bacon rashers**, back, *grilled* 4 rashers, fat trimmed	**45**		131	548	13.7	8.5	0	0	6	0.7	0	0.27	0.10	5.4	0
282	back, *grilled*	3 rashers	**45**	182	756	11.4	15.2	0	0	5	0.7	0	0.19	0.08	4.1	0
283	streaky, *grilled*	4 rashers	**40**	169	700	9.8	14.4	0	0	5	0.6	0	0.16	0.06	3.5	0
284	**Bacon steaks**, gammon grilled, *lean*	1 steak	**120**	206	871	37.7	6.2	0	0	12	1.8	0	1.20	0.32	15.6	0
285	gammon, *grilled, lean and fat*	1 steak	**120**	274	1144	35.4	14.6	0	0	11	1.7	0	1.06	0.29	14.2	0

[a] The determination of lean, separable fat and inedible proportion was described in Paul & Southgate (1977). Nutrient values for meats are given for the consumption of both lean meat and fat, and of lean meat only, with trimmable fat removed. Values are subject to review under the collaborative revision of the U.K. Food Tables by MAFF and RSC (see p. 6).

MEAT

No	Food and portion description	Portion weight		Energy		Pro	Fat	Carb	Fib	Minerals		Vitamins				
										Ca	Fe	A	B1	B2	B3	C
			g	kcal	kJ	g	g	g	g	mg	mg	µg	mg	mg	mg	mg
286	**Beef joints,** silverside															
	boiled, lean	1 serving	85	147	621	27.5	4.2	0	0	9	2.7	0	0.03	0.27	9.2	0
287	silverside, boiled, lean and fat	1 serving	85	206	860	24.3	12.1	0	0	9	2.4	0	0.03	0.23	8.0	0
288	topside, roast, lean	1 serving	85	133	560	24.8	3.7	0	0	5	2.4	0	0.07	0.30	10.8	0
289	topside, roast, lean and fat	1 serving	85	182	762	22.6	10.2	0	0	5	2.2	0	0.06	0.26	9.7	0
290	**Beef steaks,** rump, grilled, lean	1 steak	155	260	1097	44.3	9.3	0	0	11	5.4	0	0.14	0.56	19.4	0
291	rump, grilled, lean and fat	1 steak	155	338	1414	42.3	18.8	0	0	11	5.3	0	0.12	0.50	17.8	0
292	**Lamb chops,** loin, grilled, lean	2 chops	160	195	819	24.5	10.9	0	0	8	1.9	0	0.13	0.27	11.7	0
293	loin, grilled, lean and fat	2 chops	160	443	1835	29.3	36.2	0	0	11	2.4	0	0.14	0.26	12.6	0
294	**Lamb joints,** breast, roast, lean	1 serving	85	214	892	21.8	14.1	0	0	9	1.4	0	0.09	0.25	9.4	0
295	breast, roast, lean and fat	1 serving	85	349	1442	16.2	31.5	0	0	9	1.3	0	0.05	0.14	6.4	0
296	leg, roast, lean	1 serving	85	162	680	25.0	6.9	0	0	7	2.3	0	0.12	0.32	11.0	0
297	leg, roast, lean and fat	1 serving	85	226	940	22.2	15.2	0	0	7	2.1	0	0.10	0.26	9.4	0
298	**Pork chops,** loin, grilled, lean	1 chop	135	180	753	25.8	8.5	0	0	7	0.9	0	0.70	0.20	10.9	0
299	loin, grilled, lean and fat	1 chop	135	348	1449	30.0	25.4	0	0	12	1.2	0	0.69	0.22	11.5	0
300	**Pork joints,** leg, roast, lean	1 serving	85	157	660	26.1	5.9	0	0	8	1.1	0	0.72	0.30	10.5	0
301	leg, roast, lean and fat	1 serving	85	243	1012	22.9	16.8	0	0	9	1.1	0	0.55	0.23	8.5	0
302	**Pork rashers,** belly, grilled	2 rashers	200	796	3292	42.2	69.6	0	0	22	2.0	0	1.06	0.22	16.2	0

MEAT – CANNED

No	Food and portion description	Portion weight (g)	Energy kcal	Energy kJ	Pro g	Fat g	Carb g	Fib g	Minerals Ca mg	Fe mg	Vitamins A µg	B1 mg	B2 mg	B3 mg	C mg
303	**Chopped ham and pork**	2 slices **60**	162	671	8.6	14.2	0	0	8	0.7	0	0.11	0.13	3.5	0
304	**Corned beef**	2 slices **60**	130	543	16.1	7.3	0	0	8	1.7	0	0	0.14	5.4	0
305	**Ham**	2 slices **55**	66	276	10.1	2.8	0	0	5	0.7	0	0.29	0.14	3.8	0
306	**Luncheon meat**	2 slices **70**	219	909	8.8	18.8	3.9	0.1	11	0.8	0	0.05	0.08	3.2	0
	MEAT DISHES AND PRODUCTS														
307	**Beefburgers**, *fried*	2 burgers **90**	238	989	18.4	15.6	6.3	0.3	30	2.8	0	0.02	0.21	7.2	0
308	**Beef kheema**	1 serving **190**	781	3226	34.6	71.1	0.6	0.4	32	5.1	156	0.04	0.30	9.5	2
309	**Beef koftas**	5 koftas **80**	280	1164	18.3	21.7	3.1	0.4	23	2.4	50	0.02	0.13	5.6	2
310	**Beef stew**	1 serving **175**	208	872	16.8	13.1	6.3	1.1	33	2.1	467	0.07	0.18	6.7	0
311	**Bolognese sauce**	1 serving **140**	195	811	11.2	15.3	3.5	0.3	36	2.2	452	0.08	0.17	4.6	7
312	**Chilli-con-carne**	1 serving **235**	348	1455	26.1	20.0	17.4	7.5	85	7.3	113	0.21	0.33	9.2	12
313	**Cornish pasty**	1 pasty **255**	847	3539	20.4	52.0	79.3	3.1	153	3.8	0	0.26	0.15	8.4	0
314	**Frankfurters**	3 frankfurters **90**	247	1022	8.6	22.5	2.7	0.1	31	1.4	0	0.07	0.11	2.7	0
315	**Hotpot**	1 serving **195**	222	936	18.1	8.2	20.3	2.7	43	2.3	618	0.14	0.20	7.4	10
316	**Irish stew**	1 serving **255**	291	1211	12.0	17.1	23.5	2.6	28	1.3	0	0.13	0.15	7.4	10
317	**Kebab**, Indian	1 kebab **155**	553	2303	46.7	40.0	2.0	N	62	5.1	99	0.29	0.59	20.2	3
318	**Lasagne**	1 serving **230**	347	1458	15.6	19.8	28.8	0.9	225	1.8	363	0.09	0.28	5.1	Tr
319	**Meat and vegetable pie** *shortcrust pastry top and bottom*	1 slice **130**	450	1877	12.2	27.8	39.8	2.0	86	2.1	182	0.14	0.07	4.6	Tr
320	**Meat loaf**	1 slice **100**	280	1168	18.2	18.6	10.7	0.7	44	2.7	12	0.06	0.24	6.9	0

MEAT DISHES AND PRODUCTS

No	Food and portion description	Portion weight		Energy		Pro	Fat	Carb	Fib	Minerals			Vitamins			
			g	kcal	kJ	g	g	g	g	Ca mg	Fe mg	A µg	B$_1$ mg	B$_2$ mg	B$_3$ mg	C mg
321	**Minced beef, stewed with onion**															
	1 serving	**165**		317	1313	16.7	23.9	8.9	0.8	36	2.0	0	0.07	0.12	6.3	Tr
322	*made with lean mince*															
	1 serving	**165**		216	909	21.6	10.6	9.2	0.8	36	2.6	0	0.08	0.20	7.8	Tr
323	**Minced beef, stewed with vegetables**															
	1 serving	**170**		248	1035	13.1	18.4	8.3	1.4	41	1.7	583	0.07	0.10	4.9	Tr
324	*made with lean mince*															
	1 serving	**170**		184	770	17.7	8.7	9.4	1.5	44	2.4	644	0.09	0.17	6.6	Tr
325	**Moussaka**															
	1 serving	**160**		312	1298	14.9	21.4	15.7	1.3	141	2.1	66	0.10	0.24	5.4	6
326	**Mutton biriani**															
	1 serving	**225**		560	2347	18.0	30.8	55.8	N	72	2.0	304	0.11	0.23	6.8	11
327	**Pork pie**															
	1 individual pie	**150**		564	2346	14.7	40.5	37.4	1.4	71	2.1	0	0.24	0.14	5.9	0
328	**Ravioli,** *in tomato sauce, canned*															
	1 serving	**145**		102	431	4.4	3.2	14.9	1.5	23	1.2	N	0.07	0.06	2.2	Tr
329	**Salami**															
	5 slices	**55**		270	1117	10.6	24.9	1.0	0.1	6	0.6	0	0.12	0.13	4.5	0
330	**Samosa,** *meat filling*															
	2 samosas	**110**		652	2696	5.6	61.7	19.7	2.1	37	0.9	31	0.10	0.06	2.3	1
331	**Sausages, beef,** *grilled*															
	2 sausages	**90**		239	994	11.7	15.6	13.7	0.5	66	1.5	0	0	0.13	7.4	0
332	*pork, grilled*															
	2 sausages	**90**		286	1188	12.0	22.1	10.4	0.5	48	1.4	0	0.02	0.14	6.1	0
333	**Sausage roll,** *made with flaky pastry*															
	1 sausage roll	**65**		311	1294	4.7	23.5	21.5	0.8	46	0.8	81	0.07	0.03	2.1	0
334	**Saveloy**															
	1 saveloy	**75**		197	816	7.4	15.4	7.6	0.3	17	1.1	0	0.11	0.07	2.9	0
335	**Shepherds pie**															
	1 serving	**165**		196	820	12.5	10.1	14.7	0.8	25	1.8	23	0.07	0.20	5.3	3

MEAT DISHES AND PRODUCTS

No	Food and portion description	Portion weight		Energy		Pro	Fat	Carb	Fib	Minerals			Vitamins			
										Ca	Fe	A	B$_1$	B$_2$	B$_3$	C
			g	kcal	kJ	g	g	g	g	mg	mg	µg	mg	mg	mg	mg
336	**Steak and kidney pie,** *with flaky pastry top*	1 slice	**140**	400	1673	21.3	25.6	22.7	0.8	52	3.9	140	0.20	0.73	9.5	3
337	*with flaky pastry top and bottom*	1 individual pie	**165**	533	2226	15.0	35.0	42.2	1.7	87	4.1	165	0.20	0.25	5.6	0
338	**Steak and kidney pudding**	1 slice	**160**	357	1499	17.3	19.0	31.4	1.4	146	3.0	34	0.18	0.38	6.7	Tr
339	**Toad in the hole**	1 serving	**135**	383	1596	9.9	28.5	23.4	0.7	117	1.2	31	0.09	0.16	3.8	0

MILK AS A DRINK
Average of glass, cup and mug

No	Food and portion description	Portion weight		Energy		Pro	Fat	Carb	Fib	Ca	Fe	A	B$_1$	B$_2$	B$_3$	C
340	**Cows milk,** semi-skimmed	Approx 1/3 pt	**195**	90	380	6.4	3.1	9.8	0	234	0.1	45	0.08	0.35	1.7	2
341	skimmed	Approx 1/3 pt	**195**	64	273	6.4	0.2	9.8	0	234	0.1	2	0.08	0.33	1.7	2
342	whole	Approx 1/3 pt	**195**	129	536	6.2	7.6	9.4	0	224	0.1	109	0.06	0.33	1.6	2
343	**Goats milk**	Approx 1/3 pt	**195**	117	493	6.0	6.8	8.6	0	195	0.2	86	0.08	0.25	2.0	2
344	**Soya milk**	Approx 1/3 pt	**195**	62	257	5.7	3.7	1.6	Tr	25	0.8	Tr	0.12	0.53	1.2	0

MILK IN TEA AND COFFEE

No	Food and portion description	Portion weight		Energy		Pro	Fat	Carb	Fib	Ca	Fe	A	B$_1$	B$_2$	B$_3$	C
345	**Cows milk,** semi-skimmed	In cup/mug	**35**	16	68	1.2	0.6	1.8	0	42	0	8	0.01	0.06	0.3	0
346	skimmed	In cup/mug	**35**	12	49	1.2	0	1.8	0	42	0	0	0.01	0.06	0.3	0
347	whole	In cup/mug	**35**	23	96	1.1	1.4	1.7	0	40	0	20	0.01	0.06	0.3	0
348	**Goats milk**	In cup/mug	**35**	21	89	1.1	1.2	1.5	0	35	0	15	0.01	0.05	0.4	0
349	**Soya milk**	In cup/mug	**35**	11	46	1.0	0.7	0.3	Tr	5	0.1	Tr	0.02	0.09	0.2	0

34

MILK ON BREAKFAST CEREALS

On a medium portion

		Portion weight	Energy		Pro	Fat	Carb	Fib	Minerals		Vitamins				
No	Food and portion description	g	kcal	kJ	g	g	g	g	Ca mg	Fe mg	A µg	B₁ mg	B₂ mg	B₃ mg	C mg
350	Cows milk, semi-skimmed — Over cereal 115	115	53	224	3.8	1.8	5.8	0	138	0.1	26	0.05	0.21	1.0	1
351	skimmed — Over cereal 115	115	38	161	3.8	0.1	5.8	0	138	0.1	1	0.05	0.20	1.0	1
352	whole — Over cereal 115	115	76	316	3.7	4.5	5.5	0	132	0.1	64	0.03	0.20	1.0	1
353	Goats milk — Over cereal 115	115	69	291	3.6	4.0	5.1	0	115	0.1	51	0.05	0.15	1.2	1
354	Soya milk — Over cereal 115	115	37	152	3.3	2.2	0.9	Tr	15	0.5	Tr	0.07	0.31	0.7	0

MILK ON PUDDINGS

		Portion weight	Energy		Pro	Fat	Carb	Fib	Ca	Fe	A	B₁	B₂	B₃	C
355	Condensed milk, skimmed, sweetened — Over pudding 45	45	120	512	4.5	0.1	27.0	0	149	0.1	14	0.05	0.23	1.2	2
356	whole, sweetened — Over pudding 45	45	150	633	3.8	4.5	25.0	0	131	0.1	55	0.04	0.21	1.0	2
357	Evaporated milk, whole — Over pudding 30	30	45	189	2.5	2.8	2.6	0	87	0.1	37	0.02	0.13	0.7	0

NUTS AND SEEDS

Nuts: kernels only

		Portion weight	Energy		Pro	Fat	Carb	Fib	Ca	Fe	A	B₁	B₂	B₃	C
358	Almonds — 20 kernels 20	20	113	467	3.4	10.7	0.9	2.6	50	0.8	0	0.05	0.18	0.9	Tr
359	Brazil nuts — 9 kernels 30	30	186	764	3.6	18.5	1.2	2.4	54	0.8	0	0.30	0.04	1.3	0
360	Cashew nuts — 20 kernels 40	40	224	939	6.9	18.3	11.2	N	15	1.5	4	0.17	0.10	0.8	Tr
361	Hazel nuts — 30 kernels 25	25	95	393	1.9	9.0	1.7	1.4	11	0.3	0	0.10	N	0.8	0
362	Peanuts — 32 kernels 30	30	171	709	7.3	14.7	2.6	2.2	18	0.6	0	0.27	0.03	6.4	Tr
363	Sesame seeds — 1 sprinkling 15	15	88	369	4.0	8.2	1.0	N	20	1.2	1	0.11	0.01	1.9	N
364	Walnuts — 9 halves 25	25	131	542	2.7	12.9	1.3	1.2	15	0.6	0	0.08	0.03	0.8	0

No	Food and portion description	Portion weight		Energy		Pro	Fat	Carb	Fib	Minerals			Vitamins				
										Ca	Fe	A	B_1	B_2	B_3	C
			g	kcal	kJ	g	g	g	g	mg	mg	µg	mg	mg	mg	mg
OFFAL																
365	**Kidney**, lambs, *fried*	1 serving	**75**	116	488	18.5	4.7	0	0	10	9.0	120	0.42	1.73	11.2	7
366	**Liver**, lambs, *fried*	1 serving	**90**	209	873	20.6	12.6	3.5	0	11	9.0	18549	0.23	3.96	18.1	11
OFFAL DISHES AND PRODUCTS																
367	**Faggots**	2 faggots	**190**	509	2124	21.1	35.2	29.1	1.0	105	15.8	2850	0.27	0.93	9.7	0
368	**Heart casserole**	1 serving	**160**	150	630	12.5	6.7	10.6	1.8	45	3.2	381	0.27	0.54	5.9	Tr
369	**Liver and onion stew**	1 serving	**125**	211	879	14.5	14.4	6.4	0.5	20	8.5	9751	0.16	1.68	10.3	Tr
370	**Liver pâté**	1 serving	**60**	109	449	4.5	9.4	1.5	0.1	9	2.2	2905	0.06	0.55	2.3	0
371	**Liver sausage**	4 slices	**35**	186	770	7.7	16.1	2.6	0.1	16	3.8	4980	0.10	0.95	4.0	0
372	**Tongue**, *canned*	2 slices	**50**	107	442	8.0	8.3	0	0	16	1.3	0	0.02	0.20	3.2	0
373	**Tripe and onions**	1 serving	**165**	135	569	12.5	5.1	10.6	0.7	177	0.8	21	0.05	0.15	2.8	Tr
PASTA																
374	**Spaghetti**, white, *boiled*	1 serving	**150**	156	663	5.4	1.1	33.3	2.7	11	0.8	0	0.02	0.02	1.8	0
375	wholemeal, *boiled*	1 serving	**150**	170	728	7.1	1.4	34.8	6.0	17	2.1	0	0.32	0.03	3.5	0
PASTA PRODUCTS																
376	**Spaghetti**, *canned in Bolognese sauce*	1 serving	**125**	108	453	4.1	3.8	15.3	1.8	23	0.9	N	0.09	0.04	1.9	0
377	*canned in tomato sauce*	1 serving	**125**	80	341	2.4	0.5	17.6	3.5	15	0.4	N	0.09	0.01	1.3	Tr

POULTRY

No	Food and portion description	Portion weight		Energy		Pro	Fat	Carb	Fib	Minerals		A	Vitamins			
										Ca	Fe		B₁	B₂	B₃	C
			g	kcal	kJ	g	g	g	g	mg	mg	μg	mg	mg	mg	mg
378	**Chicken**, *roast, dark meat*	1 serving	**85**	132	551	19.6	5.9	0	0	8	0.9	0	0.08	0.20	8.8	0
379	*roast, light meat*	1 serving	**85**	121	509	22.5	3.4	0	0	8	0.4	0	0.07	0.12	13.0	0
380	*roast, meat and skin*	1 serving	**85**	184	767	19.2	11.9	0	0	8	0.7	0	0.05	0.14	8.9	0
381	*roast, leg portion*	1 leg portion	**190**	175	737	29.3	6.5	0	0	11	1.0	0	0.10	0.23	15.2	0
382	*roast, wing portion*	1 wing portion	**230**	170	715	28.5	6.2	0	0	12	0.9	0	0.09	0.23	14.7	0
383	**Duck**, *roast, meat*	1 serving	**85**	161	671	21.5	8.2	0	0	11	2.3	0	0.22	0.40	8.9	0
384	*roast, meat and skin*	1 serving	**85**	288	1195	16.7	24.7	0	0	10	2.3	N	N	N	N	N
385	**Turkey**, *roast, dark meat*	1 serving	**85**	126	530	23.6	3.5	0	0	10	1.2	0	0.06	0.25	10.1	0
386	*roast, light meat*	1 serving	**85**	112	474	25.3	1.2	0	0	6	0.4	0	0.06	0.12	13.3	0
387	*roast, meat and skin*	1 serving	**85**	145	609	23.8	5.5	0	0	8	0.8	0	0.06	0.18	11.6	0

POULTRY DISHES AND PRODUCTS

No	Food and portion description	Portion weight		Energy		Pro	Fat	Carb	Fib	Minerals		A	Vitamins			
			g	kcal	kJ	g	g	g	g	mg	mg	μg	mg	mg	mg	mg
388	**Chicken casserole**	1 serving	**195**	185	772	17.6	9.8	7.0	1.4	49	2.0	548	0.12	0.14	7.8	Tr
389	**Chicken curry**	1 serving	**245**	581	2406	26.2	50.7	5.4	1.0	64	4.4	551	0.07	0.22	9.8	2
390	**Chicken in white sauce**	1 serving	**155**	231	969	24.2	11.8	7.8	0.2	105	0.9	71	0.11	0.28	11.8	0
391	**Chicken pie**, *with flaky pastry top*	1 slice	**160**	472	1970	19.4	30.6	31.7	1.3	122	1.4	160	0.16	0.19	9.3	0
392	*with shortcrust pastry, top and bottom*	1 slice	**140**	473	1980	15.7	27.9	42.6	1.8	125	1.7	146	0.17	0.13	7.3	0

PUDDINGS

No	Food and portion description	Portion weight (g)	Energy kcal	Energy kJ	Pro g	Fat g	Carb g	Fib g	Minerals Ca mg	Fe mg	A µg	Vitamins B₁ mg	B₂ mg	B₃ mg	C mg
393	**Arctic roll**	1 slice **30**	60	254	1.2	2.0	10.0	0.2	27	0.2	N	0.02	0.03	0.3	0
394	**Apple,** baked	1 apple **125**	49	206	0.4	0	12.5	2.9	5	0.4	6	0.04	0.03	0.1	18
395	stewed	1 serving **120**	38	163	0.4	0	9.8	2.3	4	0.4	5	0.04	0.02	0.1	14
396	stewed with sugar	1 serving **120**	79	338	0.4	0	20.8	2.0	4	0.2	5	0.04	0.02	0.1	13
397	**Bakewell tart**	1 slice **90**	410	1713	5.7	26.7	39.2	2.4	72	1.4	153	0.08	0.14	1.8	1
398	**Blancmange**	1 serving **135**	154	648	4.2	5.0	24.6	Tr	149	0.1	70	0.04	0.19	1.1	Tr
399	**Bread and butter pudding**	1 serving **130**	208	875	8.1	10.1	22.8	0.3	169	0.9	143	0.09	0.27	2.5	1
400	**Bread pudding**	1 serving **190**	564	2379	11.2	18.2	94.4	5.7	228	3.0	209	0.19	0.23	4.0	Tr
401	**Custard tart**	1 slice **85**	238	995	4.8	14.2	24.2	0.9	94	0.8	94	0.09	0.11	1.5	Tr
402	**Eve's pudding**	1 serving **105**	253	1059	3.7	13.8	30.3	1.8	55	0.8	158	0.05	0.07	1.4	4
403	**Fruit crumble,** *made with white flour*	1 serving **120**	238	1002	2.4	8.3	40.8	2.6	59	0.7	96	0.06	0.02	1.1	4
404	*made with wholemeal flour*	1 serving **120**	232	976	3.1	8.5	38.0	3.6	38	1.1	96	0.08	0.04	1.9	4
405	**Fruit flan,** *made with shortcrust pastry*	1 slice **80**	94	397	1.1	3.5	15.4	0.6	18	0.4	44	0.05	0.02	0.6	8
406	*made with sponge*	1 slice **60**	67	286	1.7	0.9	14.0	0.4	15	0.4	22	0.04	0.04	0.7	5
407	**Fruit pie,** *made with white flour pastry crust*	1 slice **120**	223	941	2.4	9.5	34.4	2.5	58	0.6	56	0.06	0.02	1.1	6
408	*made with wholemeal flour pastry crust*	1 slice **120**	220	924	3.1	9.7	31.9	3.6	37	1.1	56	0.10	0.04	1.9	6
409	**Fruit salad,** fresh	1 serving **185**	98	416	1.1	0.2	24.6	3.9	30	0.6	19	0.09	0.06	0.7	26
410	**Gulab jamen**	1 serving **75**	269	1130	5.7	11.3	38.6	0.3	173	0.2	75	0.04	0.17	1.4	2

PUDDINGS

No	Food and portion description	Portion weight		Energy		Pro	Fat	Carb	Fib	Minerals		Vitamins				
										Ca	Fe	A	B₁	B₂	B₃	C
			g	kcal	kJ	g	g	g	g	mg	mg	µg	mg	mg	mg	mg
411	**Ice cream, dairy, vanilla**	1 serving	75	146	611	2.7	7.4	18.3	0	98	0.1	111	0.03	0.19	0.7	1
412	dairy, flavoured	1 serving	75	134	563	2.6	6.0	18.5	0	83	0.4	91	0.03	0.20	0.7	1
413	non-dairy, vanilla	1 serving	75	134	560	2.4	6.5	17.3	0	90	0.1	2	0.03	0.18	0.7	1
414	non-dairy, flavoured	1 serving	75	125	524	2.3	5.6	17.4	0	90	0.1	2	0.03	0.18	0.6	1
415	non-dairy, reduced calorie	1 serving	75	89	374	2.6	4.5	10.3	N	N	N	N	N	N	N	N
416	**Instant desserts**, *made up with*															
	semi-skimmed milk	1 sundae glass	90	85	361	2.8	4.0	13.4	0.2	90	0.1	N	0.03	0.14	0.7	1
417	*made up with skimmed milk*	1 sundae glass	90	76	320	2.8	2.9	13.4	0.2	90	0.1	N	0.03	0.14	0.7	1
418	*made up with whole milk*	1 sundae glass	90	100	420	2.8	5.7	13.3	0.2	87	0.1	N	0.03	0.13	0.7	1
419	**Jellabi**	1 serving	50	182	764	2.0	6.8	30.2	0.8	26	0.4	0	0.04	0.01	0.6	0
420	**Jelly**	1 serving	135	82	351	1.6	0	20.4	0	9	0.5	0	0	0	0	0
421	**Lemon meringue pie**	1 slice	95	303	1275	4.3	13.7	43.6	0.8	43	0.9	95	0.07	0.08	1.4	3
422	**Milk jelly**, *made up with*															
	semi-skimmed milk	1 serving	135	107	456	3.5	0.9	22.7	0	80	0.5	14	0.03	0.11	0.4	1
423	*made up with skimmed milk*	1 serving	135	99	424	3.5	Tr	22.7	0	80	0.5	Tr	0.03	0.11	0.4	1
424	*made up with whole milk*	1 serving	135	119	504	3.5	2.3	22.5	0	77	0.5	32	0.03	0.09	0.4	Tr
425	**Pancakes**, *made with skimmed milk*	2 pancakes	70	196	823	4.2	9.7	24.6	0.6	77	0.6	19	0.07	0.12	1.3	1
426	*made with whole milk*	2 pancakes	70	211	882	4.1	11.3	24.5	0.6	77	0.6	41	0.07	0.12	1.3	1
427	**Queen of puddings**	1 serving	125	266	1123	6.0	9.8	41.4	0.5	99	0.9	125	0.06	0.19	1.8	1

No	Food and portion description	Portion weight	Energy		Pro	Fat	Carb	Fib	Minerals		A	B₁	B₂	B₃	C
									Ca	Fe				Vitamins	
		g	kcal	kJ	g	g	g	g	mg	mg	µg	mg	mg	mg	mg
PUDDINGS															
428	**Rice pudding**, *made with semi-skimmed milk*	1 serving **85**	91	388	3.4	1.5	17.1	0.2	111	0.1	20	0.03	0.14	0.9	1
429	*made with skimmed milk*	1 serving **85**	79	338	3.4	0.2	17.1	0.2	111	0.1	1	0.03	0.14	0.9	1
430	*made with whole milk*	1 serving **85**	110	462	3.3	3.7	16.9	0.2	111	0.1	51	0.03	0.14	0.9	1
431	**Sevyiaan**	1 serving **125**	553	2296	6.0	34.5	55.5	2.8	88	1.1	363	0.09	0.13	1.9	Tr
432	**Sponge pudding**, *made with dried fruit*	1 serving **95**	314	1322	5.1	13.6	45.7	1.5	78	1.0	143	0.09	0.08	1.7	Tr
433	*made with jam or treacle*	1 serving **95**	316	1328	4.8	13.7	46.3	1.0	72	1.0	143	0.08	0.08	1.6	Tr
434	**Treacle tart**	1 slice **70**	258	1085	2.6	9.9	42.3	1.0	43	1.0	42	0.06	0.01	0.9	0
435	**Trifle**, *topped with dairy cream*	1 serving **175**	291	1213	4.2	16.1	34.1	0.9	119	0.5	133	0.11	0.18	1.1	7
436	*topped with Dream Topping*	1 serving **175**	259	1092	6.5	8.4	39.7	0.9	144	0.9	82	0.11	0.23	2.1	7
PULSE DISHES AND PRODUCTS															
437	**Baked beans**, *in tomato sauce, canned*	1 serving **200**	128	540	10.2	1.0	20.6	13.2	90	2.8	N	0.14	0.10	2.6	0
438	**Dahl**, chickpea	1 serving **155**	223	946	12.4	5.1	34.1	8.4	99	4.8	135	0.22	0.08	2.5	5
439	lentil	1 serving **155**	140	589	7.6	4.8	17.7	3.4	17	2.6	50	0.11	0.05	1.7	0
440	**Hummus**	1 serving **65**	120	502	4.9	8.2	7.2	N	27	1.2	N	0.10	0.03	0.7	1
441	**Pea curry**	1 serving **125**	438	1808	5.3	41.8	11.0	4.1	41	2.5	520	0.16	0.10	2.5	14
442	**Pea and potato curry**	1 serving **200**	284	1180	4.6	21.6	18.8	5.6	56	1.8	110	0.10	0.08	2.0	12
443	**Tofu**	1 serving **60**	42	175	4.4	2.5	0.4	0.2	304	0.7	N	0.04	0.01	0	0

		Portion weight	Energy		Pro	Fat	Carb	Fib	Minerals			Vitamins				
									Ca	Fe	A	B1	B2	B3	C
No	Food and portion description	g	kcal	kJ	g	g	g	g	mg	mg	µg	mg	mg	mg	mg

RICE

No	Food and portion description	g	kcal	kJ	g	g	g	g	mg	mg	µg	mg	mg	mg	mg
444	**Brown rice**, *boiled*	1 serving **165**	233	985	4.3	1.8	53.0	2.5	7	0.8	0	0.23	0.03	3.1	0
445	**White rice**, *boiled*	1 serving **165**	203	861	3.6	0.5	48.8	1.3	2	0.3	0	0.02	0.02	1.3	0

RICE DISHES

446	**Fried rice**	1 serving **190**	249	1053	4.2	6.1	47.5	2.3	13	0.6	0	0.06	0.02	1.5	2
447	**Pilau rice**	1 serving **190**	412	1718	5.1	21.9	48.8	3.0	36	1.1	0	0.21	0.06	3.0	Tr
448	**Risotto**	1 serving **190**	426	1792	5.7	17.7	65.4	2.5	46	0.6	146	0.28	0.02	4.2	Tr
449	**Savoury rice**	1 serving **190**	270	1138	5.5	6.7	50.0	2.5	48	1.0	N	0.19	0.02	3.2	0

SALADS AND RAW VEGETABLES

450	**Beansprouts**, *raw*	1 serving **85**	30	124	3.2	0.2	5.0	N	16	1.1	2	0.11	0.11	0.7	16
451	**Beetroot**, *boiled, sliced*	1 serving **40**	18	76	0.7	0	4.0	0.9	12	0.2	0	0.01	0.02	0.2	2
452	*raw, grated*	1 serving **35**	10	41	0.5	0	2.1	1.0	9	0.1	0	0.01	0.02	0.1	2
453	**Brussels sprouts**, *raw, grated*	1 serving **55**	14	61	2.2	0	1.5	2.1	18	0.4	37	0.06	0.08	0.8	50
454	**Cabbage**, red, *raw, shredded*	1 serving **55**	11	47	0.9	0	1.9	1.7	29	0.3	2	0.03	0.03	0.3	30
455	white, *raw, shredded*	1 serving **55**	12	51	1.0	Tr	2.1	1.3	24	0.2	Tr	0.03	0.03	0.3	22
456	**Carrot**, *raw, grated*	1 serving **35**	8	34	0.2	Tr	1.9	0.9	17	0.2	700	0.02	0.02	0.2	2
457	**Cauliflower**, *florets, raw*	1 serving **50**	7	28	1.0	Tr	0.8	1.0	11	0.3	3	0.05	0.05	0.6	30

SALADS AND RAW VEGETABLES

No	Food and portion description	Portion weight		Energy		Pro	Fat	Carb	Fib	Minerals		Vitamins				
										Ca	Fe	A	B₁	B₂	B₃	C
		g		kcal	kJ	g	g	g	g	mg	mg	µg	mg	mg	mg	mg
458	Celery, *sticks, raw*	1 serving	40	3	14	0.4	Tr	0.5	0.6	21	0.2	Tr	0.01	0.01	0.2	3
459	Chicory, *raw*	1 serving	45	4	17	0.4	0	0.7	N	8	0.3	0	0.02	0.02	0.3	2
460	Coleslaw	1 serving	85	68	282	1.6	4.7	5.2	1.9	36	0.4	N	0.04	0.03	0.4	27
461	Cucumber, *sliced*	1 serving	30	3	13	0.2	0	0.5	0.1	7	0.1	Tr	0.01	0.01	0.1	2
462	Endive	1 serving	40	4	19	0.7	0	0.4	0.8	18	1.1	133	0.02	0.04	0.3	5
463	Lettuce	1 serving	30	4	15	0.3	0.1	0.4	0.4	7	0.3	50	0.02	0.02	0.1	5
464	Mushrooms, *raw, sliced*	1 serving	35	5	19	0.6	0.2	0	0.8	1	0.4	0	0.04	0.14	1.6	1
465	Mustard and cress	1 serving	15	2	7	0.2	0	0.1	0.5	10	0.2	12	N	N	N	6
466	Onion, *raw, rings*	1 serving	30	7	30	0.3	Tr	1.6	0.4	9	0.1	0	0.01	0.02	0.1	3
467	Onions, spring onions	3 onions	15	5	23	0.1	Tr	1.3	0.4	21	0.2	Tr	0	0.01	0.1	4
468	Parsley, *chopped*	Sprinkling	4	1	4	0.2	0	0	0.3	13	0.3	47	0.01	0.01	0.1	6
469	Peppers, sweet, *raw, sliced*	1 serving	45	7	29	0.4	0.2	1.0	0.4	4	0.2	15	Tr	0.01	0.4	45
470	Potato salad	1 serving	105	140	586	1.4	8.7	15.1	0.5	13	0.4	N	0.04	0.02	0.7	4
471	Radishes	4 radishes	50	8	31	0.5	Tr	1.4	0.5	22	1.0	Tr	0.02	0.01	0.2	13
472	Tomato, *raw*	2 tomatoes	150	21	90	1.4	Tr	4.2	2.1	20	0.6	150	0.09	0.06	1.2	30
473	Vegetable salad, *canned*	1 serving	100	122	510	1.7	8.4	10.7	2.5	23	0.9	273	0.03	0.03	0.8	7
474	Watercress	1 serving	15	2	9	0.4	Tr	0.1	0.5	33	0.2	75	0.02	0.02	0.2	9

SAUCES – SAVOURY

No	Food and portion description	Portion weight		Energy		Pro	Fat	Carb	Fib	Minerals			Vitamins				
			g	kcal	kJ	g	g	g	g	Ca mg	Fe mg	A µg	B₁ mg	B₂ mg	B₃ mg	C mg
475	**Apple purée**	1 serving	80	26	109	0.2	0	6.6	1.5	2	0.2	3	0.02	0.02	0.1	10
476	**Bread sauce,** *made with*															
	semi-skimmed milk	1 serving	80	74	314	3.4	2.5	10.2	0.5	96	0.2	27	0.04	0.11	1.0	1
477	*made with skimmed milk*	1 serving	80	65	276	3.4	1.4	10.2	0.5	96	0.2	13	0.04	0.11	1.0	1
478	*made with whole milk*	1 serving	80	88	370	3.4	4.1	10.1	0.5	96	0.2	50	0.04	0.11	1.0	Tr
479	**Brown sauce,** *bottled*	1 serving	9	9	38	0.1	0	2.3	N	4	0.3	N	N	N	N	N
480	**Cheese sauce,** *made with*															
	semi-skimmed milk	1 serving	80	143	600	6.5	10.1	7.3	0.2	192	0.2	113	0.04	0.18	1.7	1
481	*made with skimmed milk*	1 serving	80	134	562	6.5	9.0	7.3	0.2	192	0.2	95	0.04	0.18	1.7	1
482	*made with whole milk*	1 serving	80	158	655	6.4	11.7	7.2	0.2	192	0.2	134	0.04	0.18	1.7	1
483	**French dressing**	1 serving	9	59	244	0	6.6	0	0	0	0	0	0	0	0	0
484	**Gravy,** *cornflour based*	1 serving	80	20	85	1.0	1.0	2.1	N	N	N	N	N	N	N	N
485	*meat juice based*	1 serving	80	87	366	1.5	7.5	3.7	N	N	N	N	N	N	N	N
486	**Mayonnaise**	1 serving	20	138	569	0.2	15.1	0.3	0	2	0.1	21	0	0.01	0.1	N
487	*reduced calorie*	1 serving	20	58	238	0.2	5.6	1.6	0	N	N	N	N	N	N	N
488	**Onion sauce,** *made with*															
	semi-skimmed milk	1 serving	80	69	289	2.3	4.0	6.7	0.5	74	0.2	50	0.05	0.10	0.8	2
489	*made with skimmed milk*	1 serving	80	62	259	2.3	3.2	6.7	0.5	74	0.2	38	0.05	0.10	0.8	2
490	*made with whole milk*	1 serving	80	79	331	2.2	5.2	6.6	0.5	72	0.2	67	0.05	0.10	0.7	1
491	**Salad cream**	1 serving	15	52	216	0.2	4.7	2.5	N	3	0.1	2	N	N	N	0
492	*reduced calorie*	1 serving	15	29	121	0.2	2.6	1.4	N	N	N	N	N	N	N	0

No	Food and portion description	Portion weight	Energy		Pro	Fat	Carb	Fib	Minerals		Vitamins				
									Ca	Fe	A	B₁	B₂	B₃	C
		g	kcal	kJ	g	g	g	g	mg	mg	µg	mg	mg	mg	mg
SAUCES – SAVOURY															
493	**Tomato ketchup** 1 serving	**20**	20	84	0.4	0	4.8	N	5	0.2	N	N	N	N	N
494	**Tomato sauce** 1 serving	**80**	69	287	1.9	4.1	6.5	1.4	22	0.6	188	0.06	0.04	1.1	8
495	**White sauce,** *made with* *semi-skimmed milk* 1 serving	**80**	102	431	3.4	6.2	8.9	0.2	112	0.2	72	0.04	0.15	1.0	1
496	*made with skimmed milk* 1 serving	**80**	91	384	3.4	5.0	8.9	0.2	112	0.2	54	0.04	0.15	1.0	1
497	*made with whole milk* 1 serving	**80**	120	499	3.3	8.2	8.7	0.2	104	0.2	102	0.04	0.14	0.9	Tr
SAUCES – SWEET															
498	**Custard,** *powder made up* *with semi-skimmed milk* 1 serving	**75**	71	302	2.9	1.4	12.6	Tr	105	0.1	20	0.03	0.14	0.8	1
499	*powder made up with skimmed milk* 1 serving	**75**	59	254	2.9	0.1	12.6	Tr	105	0.1	1	0.03	0.14	0.8	1
500	*powder made up with whole milk* 1 serving	**75**	88	371	2.8	3.4	12.5	Tr	98	0.1	47	0.03	0.14	0.8	1
501	**Dream Topping,** *powder made* *up with semi-skimmed milk* 1 serving	**35**	58	243	1.4	4.1	4.3	Tr	35	0	N	0.01	0.07	0.4	0
502	*powder made up with skimmed milk* 1 serving	**35**	54	228	1.4	3.7	4.3	Tr	35	0	N	0.01	0.07	0.4	0
503	*powder made up with whole milk* 1 serving	**35**	64	265	1.3	4.7	4.2	Tr	33	0	N	0.01	0.07	0.3	0
504	**Egg custard sauce** 1 serving	**75**	89	371	4.3	4.5	8.3	0	98	0.3	70	0.04	0.18	1.2	Tr
505	**White sauce,** *made with* *semi-skimmed milk* 1 serving	**75**	113	476	2.9	5.4	14.1	0.2	98	0.2	62	0.04	0.14	0.8	1
506	*made with skimmed milk* 1 serving	**75**	104	435	2.9	4.3	14.1	0.2	98	0.2	47	0.04	0.14	0.8	1
507	*made with whole milk* 1 serving	**75**	128	534	2.9	7.1	14.0	0.2	90	0.2	88	0.04	0.13	0.8	Tr

No	Food and portion description		Portion weight	Energy		Pro	Fat	Carb	Fib	Minerals		Vitamins				
										Ca	Fe	A	B$_1$	B$_2$	B$_3$	C
			g	kcal	kJ	g	g	g	g	mg	mg	µg	mg	mg	mg	mg
SAVOURY SNACKS																
508	**Chevda**, *e.g. Bombay Mix*	1 packet	100	494	2054	17.5	32.3	35.1	6.0	53	5.1	89	0.39	0.09	9.8	2
509	**Crisps**	1 packet	30	160	667	1.9	10.8	14.8	3.1	11	0.6	0	0.06	0.02	1.8	5
510	**Gherkins**	7 mini gherkins	20	3	11	0.2	N	0.5	0.2	4	0.1	0	0	0	0	0
511	**Olives**, black, *with stones*	9 olives	35	29	118	0.2	3.1	0	1.1	17	0.3	8	0	0	0	0
512	**Peanuts**, roasted and salted	1 small packet	25	143	591	6.1	12.3	2.2	1.8	15	0.5	0	0.06	0.03	5.3	0
SCONES																
513	**Scones**, *made with white flour, cheese*	1 scone	50	182	761	5.1	8.9	21.6	1.0	125	0.6	85	0.07	0.06	1.6	Tr
514	*made with white flour, fruit*	1 scone	50	158	667	3.7	4.9	26.5	1.8	75	0.8	N	0.12	0.05	1.4	Tr
515	*made with white flour, plain*	1 scone	50	181	762	3.6	7.3	26.9	1.1	90	0.7	70	0.08	0.04	1.3	Tr
516	*made with wholemeal flour, fruit*	1 scone	50	162	683	4.1	6.4	23.6	2.6	55	1.2	60	0.10	0.05	2.3	Tr
517	*made with wholemeal flour, plain*	1 scone	50	163	684	4.4	7.2	21.6	2.5	55	1.2	65	0.11	0.05	2.5	Tr
SOFT DRINKS AND FRUIT JUICES																
518	**Coca-cola**	1 glass	200	78	336	0	0	21.0	0	8	0	0	0	0	0	0
519	**Grapefruit juice**, canned, sweetened	1 glass	200	76	328	1.0	0	19.4	0	18	0.6	0	0.08	0.02	0.6	58
520	canned, unsweetened	1 glass	200	62	264	0.6	0	15.8	0	18	0.6	0	0.08	0.02	0.6	56
521	**Lemonade**	1 glass	200	42	180	0	0	11.2	0	10	0	0	0	0	0	0
522	**Lemon juice**, fresh	2 teaspoons	10	1	3	0	0	0.2	0	1	0	0	0	0	0	5

SOFT DRINKS AND FRUIT JUICES

No	Food and portion description	Portion weight		Energy		Pro	Fat	Carb	Fib	Minerals		Vitamins				
										Ca	Fe	A	B$_1$	B$_2$	B$_3$	C
			g	kcal	kJ	g	g	g	g	mg	mg	µg	mg	mg	mg	mg
523	Lime cordial, concentrated, *to make up*															
	approx. 1/3 pt	1 glass	**45**	50	216	0	0	13.4	0	4	0.1	0	0	0	0	0
524	**Lucozade**	1 glass	**200**	136	576	0	0	36.0	0	10	0.2	0	0	0	0	6
525	**Orange juice**, canned, sweetened	1 glass	**200**	102	434	1.4	0	25.6	0	18	0.6	16	0.14	0.04	0.6	62
526	canned, unsweetened	1 glass	**200**	66	286	0.8	0	17.0	0	18	1.0	16	0.14	0.04	0.6	70
527	freshly squeezed	1 small glass	**70**	27	113	0.4	0	6.6	0	8	0.2	6	0.06	0.01	0.2	35
528	**Orange squash**, concentrated, *to make up*															
	approx. 1/3 pt	1 glass	**45**	48	205	0	0	12.8	0	4	0	0	0	0	0	0
529	**Pineapple juice**, canned	1 glass	**200**	106	450	0.8	0.2	26.8	0	24	1.4	14	0.10	0.04	0.6	16
530	**Ribena**, concentrated, *to make up*															
	approx. 1/3 pt	1 glass	**45**	103	439	0	0	27.4	0	4	0.2	N	N	N	N	95
531	**Tomato juice**, canned	1 glass	**200**	32	132	1.4	0	6.8	0	20	1.0	166	0.12	0.06	1.6	40

SOUPS – CANNED

No	Food and portion description	Portion weight		Energy		Pro	Fat	Carb	Fib	Ca	Fe	A	B$_1$	B$_2$	B$_3$	C
532	**Chicken**, cream of	1 bowl	**145**	84	351	2.5	5.5	6.5	N	39	0.6	0	0.01	0.04	0.7	0
533	**Mushroom**, cream of	1 bowl	**145**	77	322	1.6	5.5	5.7	N	44	0.4	0	0	0.07	0.7	0
534	**Oxtail**	1 bowl	**145**	64	268	3.5	2.5	7.4	N	58	1.5	0	0.03	0.04	1.7	0
535	**Tomato**, cream of	1 bowl	**145**	80	334	1.2	4.8	8.6	N	25	0.6	51	0.04	0.03	0.9	0
536	**Vegetable**	1 bowl	**145**	54	231	2.2	1.0	9.7	N	25	0.9	0	0.04	0.03	0.9	0

SPREADS – SAVOURY

Spread on bread from large loaf

No	Food and portion description		Portion weight	Energy		Pro	Fat	Carb	Fib	Minerals		Vitamins				
										Ca	Fe	A	B$_1$	B$_2$	B$_3$	C
			g	kcal	kJ	g	g	g	g	mg	mg	µg	mg	mg	mg	mg
537	Bovril	Medium layer	4	7	29	1.5	0	0.1	0	2	0.6	0	0.36	0.30	3.4	0
538	Cheese spread	1 triangle	15	41	171	2.0	3.4	0.7	0	63	0	44	0.01	0.05	0.5	Tr
539	Fish paste	Medium layer	9	15	63	1.4	0.9	0.3	0	25	0.8	N	0	0.02	0.6	0
540	Marmite	Medium layer	4	7	30	1.6	0	0.1	N	4	0.1	0	0.12	0.44	2.7	0
541	Meat paste	Medium layer	9	16	65	1.4	1.0	0.3	0	8	0.2	0	0	0.02	0.6	0
542	Peanut butter	Medium layer	7	44	181	1.6	3.8	0.9	0.5	3	0.1	0	0.01	0.01	1.4	0

SPREADS – SWEET

Spread on bread from large loaf

No	Food and portion description		Portion weight	Energy		Pro	Fat	Carb	Fib	Minerals		Vitamins				
										Ca	Fe	A	B$_1$	B$_2$	B$_3$	C
			g	kcal	kJ	g	g	g	g	mg	mg	µg	mg	mg	mg	mg
543	Honey	Medium layer	10	29	123	0	0	7.6	N	1	0	0	0	0.01	0	0
544	Jam with seeds, e.g. blackberry, strawberry	Medium layer	10	26	111	0.1	0	6.9	0.1	2	0.2	0	0	0	0	1
545	Jam, stone fruit, e.g. apricot, damson	Medium layer	10	26	112	0	0	6.9	0.1	1	0.1	0	0	0	0	0
546	Lemon curd, egg based	Medium layer	10	29	122	0.3	1.4	4.1	0	2	0.1	14	0	0.01	0.1	1
547	starch based	Medium layer	10	28	120	0.1	0.5	6.3	0	1	0.1	1	0	0	0	0
548	Marmalade	Medium layer	10	26	111	0	0	7.0	0.1	4	0.1	1	0	0	0	1

No	Food and portion description	Portion weight	Energy		Pro	Fat	Carb	Fib	Minerals		Vitamins				
									Ca	Fe	A	B1	B2	B3	C
		g	kcal	kJ	g	g	g	g	mg	mg	µg	mg	mg	mg	mg
STUFFING															
549	**Sage and onion**	1 serving **60**	139	577	3.1	8.9	12.2	1.4	35	0.6	95	0.07	0.04	1.2	1
SUGAR															
550	**Honey**	1 teaspoon **7**	20	86	0	0	5.3	N	0	0	0	0	0	0	0
551	**Sugar**, brown	1 teaspoon **5**	20	84	0	0	5.2	0	3	0	N	Tr	Tr	0	0
552	white	1 teaspoon **5**	20	84	Tr	0	5.3	0	0	Tr	0	0	0	0	0
SWEETS															
553	**Boiled sweets**	Approx. ¼ lb **100**	327	1397	0	0	87.3	0	5	0.4	0	0	0	0	0
554	**Fruit gums**	1 tube **30**	52	220	0.3	0	13.4	N	108	1.3	0	0	0	0	0
555	**Fruit pastilles**	1 tube **40**	101	432	2.1	0	24.8	N	16	0.6	0	0	0	0	0
556	**Liquorice allsorts**	Approx. ¼ lb **100**	313	1333	3.9	2.2	74.1	N	63	8.1	0	0	0	0.7	0
557	**Peppermints**	1 tube **30**	118	501	0.2	0.2	30.7	0	2	0.1	0	0	0	0	0
558	**Popcorn**, candied	1 packet **100**	480	2018	2.1	20.0	77.6	N	6	0.4	200	0.06	0.04	0.5	0
559	**Toffees**	Approx. ¼ lb **100**	430	1810	2.1	17.2	71.1	0	95	1.5	0	0	0	0.4	0

VEGETABLES – COOKED

No	Food and portion description	Portion weight	Energy		Pro	Fat	Carb	Fib	Minerals		Vitamins				
									Ca	Fe	A	B₁	B₂	B₃	C
		g	kcal	kJ	g	g	g	g	mg	mg	µg	mg	mg	mg	mg
560	Artichoke, globe, boiled	1 artichoke 220	15	62	1.1	0	2.6	N	42	0.4	15	0.07	0.02	1.1	7
561	Jerusalem, boiled	1 serving 120	22	94	1.9	0	3.8	N	36	0.5	0	0.12	0	N	2
562	Asparagus, boiled	4 spears 120	11	47	2.0	0	0.7	0.8	16	0.6	50	0.06	0.05	0.8	12
563	Beans, broad, boiled	1 serving 75	36	155	3.1	0.5	5.3	2.9	16	0.8	32	0.08	0.03	2.8	11
564	butter, boiled	1 serving 75	71	304	5.3	0.2	12.8	3.5	14	1.3	0	N	N	N	0
565	French, boiled	1 serving 105	7	33	0.8	0	1.2	3.0	41	0.6	70	0.04	0.07	0.5	5
566	haricot, boiled	1 serving 105	98	416	6.9	0.5	17.4	7.0	68	2.6	0	N	N	N	0
567	red kidney, boiled	1 serving 105	98	416	6.9	0.5	17.4	7.0	68	2.6	0	N	N	N	0
568	runner, boiled	1 serving 105	20	87	2.0	0.2	2.8	3.3	23	0.7	70	0.03	0.07	0.8	5
569	Beansprouts, canned	1 serving 80	7	32	1.3	Tr	0.6	2.2	10	0.8	Tr	0.02	0.02	0.4	1
570	Broccoli, boiled	1 serving 95	17	74	2.9	0	1.5	3.5	72	1.0	396	0.06	0.19	1.1	32
571	Brussels sprouts, boiled	1 serving 115	21	86	3.2	0	2.0	3.0	29	0.6	77	0.07	0.12	1.0	46
572	Cabbage, boiled	1 serving 75	7	30	1.0	0	0.8	1.7	40	0.5	38	0.02	0.02	0.3	11
573	Carrots, boiled	1 serving 65	12	51	0.4	0	2.8	1.8	24	0.3	1300	0.03	0.03	0.3	3
574	Cauliflower, boiled	1 serving 100	9	40	1.6	0	0.8	1.6	18	0.4	5	0.06	0.06	0.8	20
575	Celery, boiled	1 serving 60	3	13	0.4	0	0.4	1.2	31	0.2	0	0.01	0.01	0.2	3
576	Leeks, boiled	1 serving 125	30	130	2.3	0	5.8	4.4	76	2.5	9	0.09	0.04	0.9	19
577	Marrow, boiled	1 serving 90	6	26	0.4	0	1.3	0.5	13	0.2	5	0	0	0.3	2
578	Mushrooms, fried	1 serving 55	116	475	1.2	12.3	0	2.0	2	0.7	0	0.04	0.19	2.4	1

VEGETABLES – COOKED

No	Food and portion description	Portion weight (g)	Energy (kcal)	Energy (kJ)	Pro (g)	Fat (g)	Carb (g)	Fib (g)	Minerals Ca (mg)	Fe (mg)	A (µg)	Vitamins B₁ (mg)	B₂ (mg)	B₃ (mg)	C (mg)
579	**Onions**, *fried*	1 serving **40**	138	570	0.7	13.3	4.0	1.6	24	0.2	0	N	N	N	N
580	**Parsnips**, *boiled*	1 serving **110**	62	262	1.4	0	14.9	2.5	40	0.6	0	0.08	0.07	1.0	11
581	**Peas**, frozen, *boiled*	1 serving **75**	31	131	4.1	0.3	0	8.1	1	1.1	38	0.18	0.05	1.8	10
582	garden, canned	1 serving **85**	40	171	3.9	0.3	6.0	4.8	20	1.4	43	0.11	0.09	2.4	7
583	marrowfat, processed	1 serving **85**	68	288	5.3	0.3	11.6	6.0	23	1.3	43	0.09	0.03	1.3	0
584	**Plantain**, green, *boiled*	1 serving **85**	104	440	0.9	0.1	26.4	4.9	8	0.3	9	0	0.01	0.4	3
585	ripe, *fried*	1 serving **80**	214	901	1.2	7.4	38.0	4.2	5	0.6	16	0.09	0.02	0.6	10
586	**Potatoes**, *boiled*	1 serving **150**	120	515	2.1	0.2	29.6	1.4	6	0.5	0	0.12	0.05	1.7	14
587	chips	Chip shop portion **265**	670	2822	10.1	28.9	98.8	N	37	2.4	0	0.27	0.11	5.6	27
588	jacket	1 potato **140**	147	627	3.6	0.1	35.0	3.2	13	1.1	0	0.14	0.06	2.5	14
589	mashed	1 serving **170**	202	848	2.6	8.5	30.6	1.4	20	0.5	0	0.14	0.07	2.0	14
590	roast	1 serving **130**	204	861	3.6	6.2	35.5	N	13	0.9	0	0.13	0.05	2.5	13
591	sweet, *boiled*	1 serving **150**	128	545	1.7	0.9	30.2	3.2	32	0.9	1001	0.12	0.06	1.4	23
592	**Spinach**, *boiled*	1 serving **130**	39	166	6.6	0.7	1.8	7.4	780	5.2	1300	0.09	0.20	2.3	33
593	**Spring greens**, *boiled*	1 serving **75**	8	32	1.3	0	0.7	2.6	65	1.0	500	0.05	0.15	0.6	23
594	**Swede**, *boiled*	1 serving **120**	22	91	1.1	0	4.6	3.0	50	0.4	0	0.05	0.04	1.2	20
595	**Sweetcorn kernels**, canned	1 serving **70**	53	228	2.0	0.4	11.3	3.6	2	0.4	25	0.04	0.06	1.1	4

		Portion weight	Energy		Pro	Fat	Carb	Fib	Minerals		Vitamins				
No	Food and portion description								Ca	Fe	A	B1	B2	B3	C
		g	kcal	kJ	g	g	g	g	mg	mg	µg	mg	mg	mg	mg

VEGETABLES – COOKED

No	Food and portion description	weight g	kcal	kJ	Pro	Fat	Carb	Fib	Ca	Fe	A	B1	B2	B3	C
596	Tomatoes, canned	1 serving 140	17	71	1.5	Tr	2.8	1.1	13	1.3	116	0.08	0.04	1.1	25
597	Tomatoes, fried	2 tomatoes 140	97	403	1.4	8.3	4.6	3.8	21	0.7	N	N	N	N	14
598	Turnips, boiled	1 serving 120	17	72	0.8	0.4	2.8	2.4	66	0.5	0	0.04	0.05	0.7	20
599	Yam, boiled	1 serving 130	155	660	2.1	0.1	38.7	4.6	12	0.4	3	0.07	0.01	1.0	3

VEGETABLE DISHES AND PRODUCTS

No	Food and portion description	weight g	kcal	kJ	Pro	Fat	Carb	Fib	Ca	Fe	A	B1	B2	B3	C
600	Bubble and squeak	1 serving 145	186	780	2.3	12.9	16.2	2.8	30	0.6	36	0.09	0.04	1.2	10
601	Cauliflower bhajia	1 serving 180	193	801	4.7	15.7	9.0	4.7	104	5.2	392	0.07	0.14	1.8	27
602	Cauliflower cheese	1 serving 165	173	723	9.7	11.4	8.4	2.3	198	1.0	125	0.17	0.17	3.1	12
603	Cauliflower with white sauce	1 serving 165	92	388	4.1	5.6	6.9	2.1	96	0.7	64	0.10	0.15	1.5	21
604	Mixed vegetables, boiled	1 serving 75	32	134	1.6	0.2	6.3	2.7	20	0.5	1001	0.06	0.05	1.0	6
605	Okra curry	1 serving 140	363	1499	3.6	34.7	9.8	N	109	2.2	456	0.14	0.13	1.4	34
606	Pakoras	1 serving 60	108	455	4.7	4.9	12.2	3.4	56	2.0	78	0.09	0.03	1.0	3
607	Potato curry	1 serving 175	292	1229	4.0	15.8	36.1	3.7	47	2.1	200	0.11	0.04	1.8	14
608	Ratatouille	1 serving 250	190	790	2.3	16.0	9.8	5.3	43	1.0	78	0.10	0.08	2.3	13
609	Red cabbage, sweet and sour	1 serving 105	61	254	1.4	2.7	8.1	3.0	43	0.5	34	0.03	0.03	0.4	14
610	Samosa, vegetable filling	2 samosas 110	519	2149	3.4	46.0	24.5	2.6	35	0.9	34	0.13	0.02	1.3	4
611	Vegetable curry	1 serving 220	398	1657	4.0	33.4	22.0	N	86	1.8	1569	0.15	0.11	2.2	26

YOGURT

No	Food and portion description	Portion weight		Energy		Pro	Fat	Carb	Fib	Minerals		Vitamins				
										Ca	Fe	A	B1	B2	B3	C
			g	kcal	kJ	g	g	g	g	mg	mg	µg	mg	mg	mg	mg
612	**Greek yogurt**, cows milk	1 small carton	150	173	716	9.6	13.7	3.0	0	225	0.5	182	0.05	0.54	2.4	Tr
613	ewes milk	1 small carton	150	159	663	6.6	11.3	8.4	0	225	Tr	132	0.08	0.50	1.9	Tr
614	**Low calorie yogurt**															
	assorted flavours	1 small carton	150	62	266	6.5	0.3	9.0	N	195	0.2	Tr	0.06	0.44	1.7	2
615	**Low fat yogurt**, flavoured															
		1 small carton	150	135	576	5.7	1.4	26.9	Tr	225	0.2	15	0.08	0.32	1.5	2
616	fruit	1 small carton	150	135	573	6.2	1.1	26.9	N	225	0.2	17	0.08	0.32	1.7	2
617	muesli/nut	1 small carton	150	168	711	7.5	3.3	28.8	N	255	0.3	18	0.11	0.36	2.1	2
618	plain	1 small carton	150	84	354	7.7	1.2	11.3	Tr	285	0.2	14	0.08	0.38	2.0	2
619	**Soya yogurt**	1 small carton	150	108	458	7.5	6.3	5.9	N	N	N	36	N	N	N	0
620	**Whole milk yogurt**, fruit	1 small carton	150	158	662	7.7	4.2	23.6	N	240	Tr	63	0.09	0.45	2.1	2
621	goats milk	1 small carton	150	95	395	5.3	5.7	5.9	0	180	0.3	N	0.06	0.26	1.7	2
622	'organic'	1 small carton	150	84	354	6.5	4.4	8.7	0	210	0.2	45	0.06	0.30	1.7	2
623	plain	1 small carton	150	119	500	8.6	4.5	11.7	N	300	0.2	48	0.09	0.41	2.3	2

ALCOHOLIC DRINKS [a] – based on pub measures

No	Food and portion description	Portion volume (ml)	Energy kcal	Energy kJ	Pro g	Fat g	Carb g	Fib g	Alc g	Minerals Ca mg	Fe mg	A µg	Vitamins B₁ mg	B₂ mg	B₃ mg	C mg
624	**Beer**, brown ale, bottled	1 bottle **275**	77	322	0.8	0	8.3	0	6.1	19	0.1	0	0	0.06	1.1	0
625	bitter, draught	1 pint **568**	182	750	1.7	0	13.1	0	17.6	62	0.1	0	0	0.23	3.4	0
626	mild, draught	1 pint **568**	142	591	1.1	0	9.1	0	14.8	57	0.1	0	0	0.17	2.3	0
627	keg, bitter	1 pint **568**	176	733	1.7	0	13.1	0	17.0	45	0.1	0	0	0.17	2.6	0
628	lager, bottled	1 bottle **275**	80	330	0.6	0	4.1	0	8.8	11	0	0	0	0.06	1.5	0
629	pale ale, bottled	1 bottle **275**	88	366	0.8	0	5.5	0	9.1	25	0.1	0	0	0.06	1.4	0
630	stout, bottled	1 bottle **275**	102	429	0.8	0	11.6	0	8.0	22	0.1	0	0	0.08	1.2	0
631	**Cider**, dry	1 pint **568**	204	863	0	0	14.8	0	21.6	45	2.8	0	0	0	0.1	0
632	sweet	1 pint **568**	239	1000	0	0	24.4	0	21.0	45	2.8	0	0	0	0.1	0
633	**Wine**, red	4 fl oz **114**	78	324	0.2	0	0.3	0	10.8	8	1.0	0	0	0.02	0.1	0
634	rosé	4 fl oz **114**	81	335	0.1	0	2.9	0	9.9	14	1.1	0	0	0.01	0.1	0
635	white, dry	4 fl oz **114**	75	314	0.1	0	0.7	0	10.4	10	0.6	0	0	0.01	0.1	0
636	white, sparkling	4 fl oz **114**	87	359	0.3	0	1.6	0	11.3	3	0.6	0	0	0.01	0.1	0
637	white, sweet	4 fl oz **114**	107	449	0.2	0	6.7	0	11.6	16	0.7	0	0	0.01	0.1	0
638	**Wine, fortified**, Port	⅓ gill **47**	74	308	0	0	5.6	0	7.5	2	0.2	0	0	0	0	0
639	Sherry, dry	⅓ gill **47**	55	226	0.1	0	0.7	0	7.4	3	0.2	0	0	0	0	0
640	Sherry, medium	⅓ gill **47**	55	230	0	0	1.7	0	7.0	4	0.2	0	0	0	0	0
641	Sherry, sweet	⅓ gill **47**	64	267	0.1	0	3.2	0	7.3	3	0.2	0	0	0	0	0
642	**Vermouth**, dry	⅓ gill **47**	55	232	0	0	2.6	0	6.5	3	0.2	0	0	0	0	0
643	sweet	⅓ gill **47**	71	297	0	0	7.5	0	6.1	3	0.2	0	0	0	0	0
644	**Liqueurs**, *e.g. Cherry Brandy, Curaçao*	⅙ gill **24**	75	313	0	0	6.8	0	5.8	N	N	0	0	0	0	0
645	*e.g. Brandy, Whisky*	⅙ gill **24**	53	221	0	0	0	0	7.6	0	0	0	0	0	0	0

[a] NACNE recommends that alcohol should contribute a maximum of 5% of the energy in the diet.   1g alcohol = 7kcal.
8g alcohol = 1 unit of alcohol.   Recommended maximum intakes: Women 14 – 35 units/week; Men 21 – 50 units/week.

Appendices

Appendices

Appendix 1: *VITAMIN B₆, TOTAL FOLATE and VITAMIN E*

Foods that are useful sources of vitamin B_6 (0.10mg/100mg or more), total folate (20µg/100g or more) and vitamin E (0.30mg/100g or more)

		Vitamin B_6 (mg)	Total folate (µg)	Vitamin E (mg)
Beverages:	Cocoa powder	-	38	0.4
Breakfast cereals:				
	All-Bran	1.80	250	2.20
	Muesli, Swiss style	1.60	140	3.20
	Shredded wheat	0.24	29	1.20
Cheese dishes:				
	Cauliflower cheese	0.16	7	0.41
	Pizza, cheese and tomato	0.09	23	1.27
Eggs:	Eggs, boiled	0.12	39	1.11
	Omelette	0.09	30	1.13
Grains, flours:	Bran, wheat	1.38	260	2.60
	Flour, wholemeal	0.50	57	1.40
	Flour, white, plain	0.15	22	0.30
Liver:	Lamb, fried	0.49	240	0.32
	Ox, stewed	0.52	290	0.44
	Pig, stewed	0.64	110	-
Nuts:	Almonds	0.10	50	-
	Peanuts	0.50	110	8.1
	Walnuts	0.73	66	0.8
Vegetables:	Cabbage, spring, boiled	0.10	50	-
	Laverbread	-	47	1.1
Vegetable oils:	Wheatgerm oil	-	-	133
	Sunflower seed oil	-	-	48.7
Miscellaneous:	Bovril	0.53	1040	-
	Marmite	1.3	1010	-

(Holland *et al.*, 1988, 1989; Paul & Southgate, 1978)

Appendix 2: **VITAMIN B₁₂** content of some vegan foods (μg/100g food)

Yeast extracts:	Tastex, Community, Meridian	50.0
	Vecon stock cube	13.3
	Marmite	8.3
Textured vegetable protein (TVP):	Solus	8.0
	Itona	5.0
	Granose	5.6
	Protoveg	5.7
Seaweed	Dried (various)	1.6 – 100
	Laverbread	1.6
Spirulina, dried		25.5 – 74
Soya bean tempeh		0.02 – 6.3
Soya milk:	Plamil – concentrated	3.2
	Plamil – on dilution	1.6
Margarines:	Mathews, Hawthorne Vale, Suma	5.0

(Modified from Langley, 1988)

Appendix 3: **VITAMIN D**

Foods that are useful sources of vitamin D (μg/100g food)

Cheese dishes:	Macaroni cheese	0.36
	Quiche, cheese and egg	0.93
Eggs:	Hens', boiled	1.75
Fats:	Butter	0.76
	Margarine	7.94
	Low fat spread	8.00
Liver:	Lamb, fried	0.50
	Ox, stewed	1.13

(Holland *et al.*, 1989; Paul & Southgate, 1978)

Appendix 4: *MAGNESIUM*

Foods that are good sources of magnesium (20mg/100g or more)

Bacon

Gammon rashers, grilled, lean only	33
Bacon rashers, fried average, lean only	25

Beef

Rump steak, fried, lean only	25
Sirloin, roast, lean only	22

Beverages

Cocoa powder	520
Coffee, instant, powder/granules	390

Breakfast cereals

All-Bran	370
Muesli, Swiss style	85
Shredded Wheat	130

Chicken

Roast meat only	24

Confectionary

Chocolate, plain	55
Chocolate, milk	100
Fruit gums	110

Grains, flours

Bran, wheat	520
Flour, wholemeal	120
Flour, white, plain	20

Fish

Cod, baked	26
Haddock, fried	31
Herring, fried	35
Mackerel	35
Plaice, fried in batter	21
Sardines, canned in tomato sauce	51
Shrimps, boiled	110
Oysters	42

Fruit

Raspberries	22

Liver

Lamb, fried	22
Pig, stewed	22

Meat products

Beefburgers, fried	23

Nuts (kernels only)

Almonds	260
Brazil nuts	410
Peanuts	180
Walnuts	130

Sugars

Treacle, black	140

Vegetables

Laverbread	31
Peas, frozen, raw	23
Potatoes, baked	29
Potatoes, chipped	43

Miscellaneous

Bovril	61
Curry powder	280
Marmite	180

(Holland *et al.*, 1988; Paul & Southgate, 1978)

Appendix 5: *ZINC*

Foods that are good sources of zinc (1.5mg/100g or more)

Bacon

Gammon rashers, grilled, lean only	3.5
Bacon rashers, fried, average, lean only	3.6
Bacon rashers, grilled average, lean only	3.7

Beef

Rump steak, fried, lean only	5.9
Sirloin, roast, lean only	5.5

Beverages

Cocoa powder	6.9

Breakfast cereals

All-Bran	8.4
Muesli, Swiss style	2.5
Shredded Wheat	2.3

Chicken

Roast meat only	1.5

Eggs

Eggs, boiled	1.5

Fish

Sardines, canned in tomato sauce	2.7
Shrimps, boiled	5.3
Oysters	45.0
	(6-100)

Grains, flours

Bran, wheat	16.2
Flour, wholemeal	2.9

Liver

Lamb, fried	4.4
Ox, stewed	4.3
Pig, stewed	8.2

Meat products

Beefburgers, fried	4.2

Nuts (kernels only)

Almonds	3.1
Brazil nuts	4.2
Peanuts	3.0
Walnuts	3.0

Miscellaneous

Bovril	1.8
Marmite	2.1

(Holland *et al.*, 1988, 1989; Paul & Southgate, 1978)

Appendix 6: *IODINE*

Iodine content (expressed as µg/100g) of selected foods in Britain

Cereals products

Bread: White	6
Hovis	22
Currant loaf	29
Rice, boiled	5
Crispbread	15
Cornflakes	10
Biscuits: Assorted	15, 540[a]
Cakes and pastries:	
Fruit cakes	153[a]
Bakewell slices	240[a]

Confectionery

Sweets, assorted	480[a]
Chocolates, assorted, plain	142[a]
assorted, milk	94[a]

Eggs

Eggs	53

Fats

Butter	38
Margarine, hard	23
soft	27

Fish

Cod fillet	74
Haddock, grilled	210
Plaice, grilled	13
Kippers, grilled	86

Fruit

Oranges	2
Bananas	8
Fruit juices	5

Meat

Beef stewing steak	6
Lamb, New Zealand	2
English	9
Pork, leg	3
Bacon	11
Chicken, broiler, roast	5

Milk and Milk Products

Liquid milk, summer	7
winter	37
Yogurt, plain, low fat	63
Cheese, Cheddar	39

Vegetables

Potatoes	3
Cabbage	2
Carrots	2
Peas, canned	13
Baked beans	3

Miscellaneous

Salt	44
Salt, iodised	3100
Vecon stock cube (vegetable stock cube)	155

(Modified from Wenlock *et al.*, 1982)

[a] These high values result from the use of the red colorant erythrosine; a small amount of the iodine from erythrosine may be absorbed by the body

REFERENCES

Bingham, S. and Day, K. (1987) Average portion weights of food consumed by a randomly selected British population sample. *Hum. Nutr.;Appl. Nutr.*, **41A**, 258-264.

Bingham, S., McNeil, N.I. and Cummings, J.H. (1981) The diet of individuals: a study of a randomly shown cross section of British adults in a Cambridgeshire village. *Br.J.Nutr.*, **45**, 23-35.

Committee on Dietary Allowances, Food and Nutrition Board, National Research Council (1980) *Recommended dietary allowances; Ninth revised edition.* National Academy of Sciences, Washington, D.C.

Committee on Medical Aspects of Food Policy (1984) *Diet and cardiovascular disease.* HMSO, London.

Crawley, H. (1988) *Food portion sizes.* HMSO, London.

Davies, J. and Dickerson, J. (1989) *Food facts and figures.* Faber & Faber, London.

Department of Health and Social Security (1979) *Recommended daily amounts of food energy and nutrients for groups of people in the United Kingdom.* Report on Health and Social Subjects No. 15, HMSO, London.

Duthie, J. (1988) *Nutrition and the Nurse.* Thesis, University of Surrey.

Holland, B., Unwin, I.D. and Buss, D.H. (1988) *Cereals and cereal products. The third supplement to McCance and Widdowson's 'The Composition of Foods' (4th edition).* Royal Society of Chemistry/ MAFF, Nottingham.

Holland, B., Unwin, I.D. and Buss, D.H. (1989) *Milk products and eggs. The fourth supplement to McCance and Widdowson's 'The Composition of Foods' (4th edition).* Royal Society of Chemistry/ MAFF, Cambridge.

Lambert, J.P. and Dickerson, J.W.T. (1989) A survey of high-fibre diet sheets used in the treatment of irritable bowel syndrome in Great Britain. *J. Hum. Nutr. Dietetics*, **2**, 429-435.

Langley, G. (1988) *Vegan Nutrition.* The Vegan Society, Oxford.

National Advisory Committee on Nutrition Education (1983). *Proposals for nutritional guidelines for healthy education in Britain.* Health Education Council, London.

Paul, A.A. and Southgate, D.A.T. (1977) A study on the composition of retail meat: dissection into lean, separable fat and inedible portion. *J. Hum. Nutr.*, **31**, 259-272.

Paul, A.A. and Southgate, D.A.T. (1978) *McCance and Widdowson's 'The Composition of Foods', Fourth revised edition.* MRC Special Report No. 297, HMSO, London.

Paul, A.A. and Southgate, D.A.T. (1979) McCance and Widdowson's 'The Composition of Foods': dietary fibre in egg, meat and fish dishes. *J. Hum. Nutr.*, **33**, 335-336.

Tan, S.P., Wenlock, R.W. and Buss, D.H. (1985) *Immigrant foods. The second supplement to McCance and Widdowson's 'The Composition of Foods' (4th edition).* HMSO, London.

Wenlock, R.W., Buss, D.H., Moxon, R.E. and Bunton, N.G. (1982) Trace nutrients 4. Iodine in British food. *Br.J.Nutr.*, **47**, 381-390.

Widdowson, E.M. and McCance, R.A. (1943) *Food tables, their scope and limitations.* Lancet, *i*, 230-232.

Wiles, S.J., Nettleton, P.A., Black, A.E. and Paul, A.A. (1980) The nutrient composition of some cooked dishes eaten in Britain: a supplementary food composition table. *J. Hum. Nutr.*, **34**, 189-223.

FOOD INDEX

- Food portions are indexed by their food number and as far as possible with sufficient detail to enable the correct code number to be assigned from this listing. However it may be necessary to refer to the main tables to check that the portion details are appropriate.

Bovril, spread on bread	537	Cauliflower, boiled	574
Bran	53	Cauliflower, florets, raw	457
Bran Buds	54	Celery, boiled	575
Bran Flakes	55	Celery, sticks, raw	458
Bran muffins	108	Chapatis, made with fat	39
Brazil nuts	359	Chapatis, made without fat	40
Bread and butter pudding	399	Cheddar cheese	120
Bread pudding	400	Cheddar-type cheese, reduced fat	121
Bread rolls, brown bap	36	Cheese and potato pie	150
Bread rolls, white bap	37	Cheese omelette	151
Bread rolls, wholemeal bap	38	Cheese pudding	152
Bread sauce, made with semi-skimmed milk	476	Cheese quiche, made with white flour pastry	156
Bread sauce, made with skimmed milk	477	Cheese quiche, made with wholemeal flour pastry	157
Bread sauce, made with whole milk	478	Cheese sauce, made with semi-skimmed milk	480
Bread, brown, from large loaf, medium sliced	33	Cheese sauce, made with skimmed milk	481
Bread, white, from large loaf, medium sliced	34	Cheese sauce, made with whole milk	482
Bread, wholemeal, from large loaf, medium sliced	35	Cheese soufflé	153
Brie cheese	117	Cheese spread	538
Broad beans, boiled	563	Cheesecake, frozen	83
Broccoli, boiled	570	Chelsea bun	47
Brown ale beer, bottled	624	Cherries	237
Brown bread, from large loaf, medium sliced	33	Cheshire cheese	122
Brown rice, boiled	444	Cheshire-type cheese, reduced fat	123
Brown sauce, bottled	479	Chevda, e.g. Bombay Mix	508
Brown sugar	551	Chicken casserole	388
Brussels sprouts, boiled	571	Chicken curry	389
Brussels sprouts, raw, grated	453	Chicken in white sauce	390
Bubble and squeak	600	Chicken pie, with flaky pastry top	391
Butter, on bread/roll	191	Chicken pie, with shortcrust pastry, top and bottom	392
Butter, on crackers, etc.	196	Chicken soup, cream of, canned	532
Butter, with jacket potato	201	Chicken, roast, dark meat	378
Butter beans, boiled	564	Chicken, roast, leg portion	381
		Chicken, roast, light meat	379
Cabbage, boiled	572	Chicken, roast, meat and skin	380
Cabbage, red, raw, shredded	454	Chicken, roast, wing portion	382
Cabbage, white, raw, shredded	455	Chickpea dahl	438
Caerphilly cheese	118	Chicory, raw	459
Camembert cheese	119	Chilli-con-carne	312
Candied popcorn	558	Chips, potato	587
Cantaloupe melon	245	Chocolate éclair	85
Carrot, raw, grated	456	Chocolate biscuits, e.g. Club, Penguin	24
Carrots, boiled	573	Chocolate cake, with butter icing	84
Cashew nuts	360	Chocolate digestive biscuits	25
Cauliflower bhajia	601	Chocolate mini roll Swiss roll	115
Cauliflower cheese, as main dish	149	Chocolate, fancy/filled	165
Cauliflower cheese, as vegetable	602	Chocolate, milk	166
Cauliflower with white sauce	603	Chocolate, plain	167
		Chopped ham and pork	303

Cider, dry	631	Croissant	88
Cider, sweet	632	Crumpets/pikelets, toasted	1
Cobblers, made with white flour	173	Crunchy Nut Corn Flakes	58
Cobblers, made with wholemeal flour	174	Cucumber, sliced	461
Coca-cola	518	Currant bread	48
Coco Pops	56	Currant bun	49
Cocoa, powder	8	Currants	271
Cod, in batter, fried	206	Curry, chicken	389
Cod, roe, in crumbs, fried	207	Curry, fish, made with haddock	226
Cod, steaks, grilled	208	Curry, fish, made with herring	227
Coffee, ground/infused	9	Curry, okra	605
Coffee, instant, powder/granules	10	Curry, pea	441
Coleslaw	460	Curry, pea and potato	442
Condensed milk, skimmed, sweetened, on puddings	355	Curry, potato	607
		Curry, vegetable	611
Condensed milk, whole, sweetened, on puddings	356	Custard tart, individual	89
		Custard tart, sliced	401
Corn Flakes	57	Custard, powder made up with semi-skimmed milk	498
Corned beef	304		
Cornish pasty	313	Custard, powder made up with skimmed milk	499
Cottage cheese	124		
Cottage cheese, reduced fat	125	Custard, powder made up with whole milk	500
Cows milk Greek yogurt	612		
Cows milk, semi-skimmed, as drink	340		
Cows milk, semi-skimmed, in tea/coffee	345	Dahl, chickpea	438
Cows milk, semi-skimmed, on cereals	350	Dahl, lentil	439
Cows milk, skimmed, as drink	341	Dairy ice cream, flavoured	412
Cows milk, skimmed, in tea/coffee	346	Dairy ice cream, vanilla	411
Cows milk, skimmed, on cereals	351	Dairy/fat spread, on bread/roll	192
Cows milk, whole, as drink	342	Dairy/fat spread, on crackers, etc.	197
Cows milk, whole, in tea/coffee	347	Dairy/fat spread, with jacket potato	202
Cows milk, whole, on cereals	352	Danish blue cheese	127
Crab, white meat, canned	209	Danish pastry	90
Crackers, cream	17	Dates	272
Crackers, wholemeal	18	Derby cheese	128
Cracotte type crispbread	19	Dessert plums	254
Cream cheese	126	Digestive biscuits, chocolate	25
Cream crackers	17	Digestive biscuits, plain	26
Cream horn	86	Double cream, in drinks/soups	176
Cream of chicken soup, canned	532	Double cream, on puddings	179
Cream of mushroom soup, canned	533	Double Gloucester cheese	129
Cream of tomato soup, canned	535	Doughnuts, jam filled	91
Cream, double, in drinks/soups	176	Doughnuts, ring	92
Cream, double, on puddings	179	Doughnuts, ring, iced	93
Cream, single, in drinks/soups	177	Dream Topping, powder made up with semi-skimmed milk	501
Cream, single, on puddings	180		
Cream, whipping, in drinks/soups	178	Dream Topping, powder made up with skimmed milk	502
Cream, whipping, on puddings	181		
Crispbread, cracotte type	19	Dream Topping, powder made up with whole milk	503
Crispbread, rye	20		
Crispie cake	87	Drinking chocolate, powder	11
Crisps	509	Dry cider	631

Dry vermouth	642
Duck, roast, meat	383
Duck, roast, meat and skin	384
Dumplings	175
Eccles cake	94
Edam cheese	130
Edam-type cheese, reduced fat	131
Egg and bacon pie	185
Egg based lemon curd	546
Egg custard sauce	504
Egg fried rice	186
Egg, boiled	182
Egg, fried	183
Egg, poached	184
Egg, Scotch	188
Egg, scrambled	189
Emmental cheese	132
Endive	462
Evaporated milk, whole, on puddings	357
Eve's pudding	402
Ewes milk Greek yogurt	613
Faggots	367
Fancy/filled chocolate	165
Feta cheese	133
Fig, green	238
Figs	273
Fish cakes, fried	225
Fish curry, made with haddock	226
Fish curry, made with herring	227
Fish fingers, fried	228
Fish paste, spread on bread	539
Fish pie	229
Flapjack	95
Fondant fancy	96
Frankfurters	314
French beans, boiled	565
French dressing	483
Fresh fruit salad	409
Fresh lemon juice	522
Freshly squeezed orange juice	527
Fried egg	183
Fried fish cakes	225
Fried fish fingers	228
Fried papadums	42
Fried rice	446
Fried rice, egg	186
Fromage frais	134
Frosties	59
Frozen peas, boiled	581
Fruit 'n' Fibre	60

Fruit cake, made with white flour	97
Fruit cake, made with wholemeal flour	98
Fruit crumble, made with white flour	403
Fruit crumble, made with wholemeal flour	404
Fruit flan, made with shortcrust pastry	405
Fruit flan, made with sponge	406
Fruit gums	554
Fruit low fat yogurt	616
Fruit pastilles	555
Fruit pie	99
Fruit pie, made with white flour pastry crust	407
Fruit pie, made with wholemeal flour pastry crust	408
Fruit salad, canned in syrup	259
Fruit salad, fresh	409
Fruit whole milk yogurt	620
Gammon bacon joints, boiled, lean	279
Gammon bacon joints, boiled, lean and fat	280
Gammon bacon steaks, grilled, lean	284
Gammon bacon steaks, grilled, lean and fat	285
Garden peas, canned	582
Gateau	100
Gherkins	510
Ginger nuts	27
Gingerbread	101
Globe artichoke, boiled	560
Goats milk whole milk yogurt	621
Goats milk, as drink	343
Goats milk, in tea/coffee	348
Goats milk, on cereals	353
Gooseberries	239
Gouda cheese	135
Grapefruit	242
Grapefruit, canned in syrup	260
Grapefruit juice, canned, sweetened	519
Grapefruit juice, canned, unsweetened	520
Grapenuts	61
Grapes, black	240
Grapes, white	241
Gravy, cornflour based	484
Gravy, meat juice based	485
Greek yogurt, cows milk	612
Greek yogurt, ewes milk	613
Green fig	238
Green plantain, boiled	584
Gruyere cheese	136
Guavas, canned in syrup	261

Gulab jamen	410
Ham	305
Haricot beans, boiled	566
Hazel nuts	361
Heart casserole	368
Herring, fillets in oatmeal, fried	210
Honey	550
Honey, spread on bread	543
Honeydew melon	246
Horlick's, powder	12
Hot cross bun	50
Hotpot	315
Hummus	440
Ice cream, dairy, flavoured	412
Ice cream, dairy, vanilla	411
Ice cream, non-dairy, flavoured	414
Ice cream, non-dairy, reduced calorie	415
Ice cream, non-dairy, vanilla	413
Indian kebab	317
Instant coffee, powder/granules	10
Instant desserts, made up with semi-skimmed milk	416
Instant desserts, made up with skimmed milk	417
Instant desserts, made up with whole milk	418
Irish stew	316
Jacket potatoes	588
Jaffa cake	102
Jam tarts, made with white flour pastry	103
Jam tarts, made with wholemeal flour pastry	104
Jam with seeds, e.g. blackberry, strawberry, spread on bread	544
Jam, stone fruit, e.g. apricot, damson, spread on bread	545
Jellabi	419
Jelly	420
Jerusalem artichoke, boiled	561
Kebab, Indian	317
Kedgeree	230
Keg beer, bitter	627
Kheema, beef	308
Kidney beans, red, boiled	567
Kidney, lambs, fried	365
Kipper, fillets, baked	211
Koftas, beef	309

Lager beer, bottled	628
Lamb chops, loin, grilled, lean	292
Lamb chops, loin, grilled, lean and fat	293
Lamb joints, breast, roast, lean	294
Lamb joints, breast, roast, lean and fat	295
Lamb joints, leg, roast, lean	296
Lamb joints, leg, roast, lean and fat	297
Lambs kidney, fried	365
Lambs liver, fried	366
Lancashire cheese	137
Lasagne	318
Leeks, boiled	576
Leg lamb joints, roast, lean	296
Leg lamb joints, roast, lean and fat	297
Leg pork joints, roast, lean	300
Leg pork joints, roast, lean and fat	301
Leicester cheese	138
Lemon	243
Lemon curd, egg based, spread on bread	546
Lemon curd, starch based, spread on bread	547
Lemon juice, fresh	522
Lemon meringue pie	421
Lemonade	521
Lentil dahl	439
Lettuce	463
Lime cordial, concentrated, to make up approx. 1/3 pt	523
Liqueurs, e.g. Brandy, Whisky	645
Liqueurs, e.g. Cherry Brandy, Curaçao	644
Liquorice allsorts	556
Liver and onion stew	369
Liver pâté	370
Liver sausage	371
Liver, lambs, fried	366
Low calorie yogurt, assorted flavours	614
Low fat spread, on bread/roll	193
Low fat spread, on crackers, etc.	198
Low fat spread, with jacket potato	203
Low fat yogurt, flavoured	615
Low fat yogurt, fruit	616
Low fat yogurt, muesli/nut	617
Low fat yogurt, plain	618
Lucozade	524
Luncheon meat	306
Lychees, canned in syrup	262
Lymeswold cheese	139
Macaroni cheese	154
Mackerel, fillets, fried	212

Madeira cake	105
Malt loaf	51
Mandarin oranges, canned in syrup	263
Mango	244
Mango, canned in syrup	264
Margarine, on bread/roll	194
Margarine, on crackers, etc.	199
Margarine, with jacket potato	204
Marmalade, spread on bread	548
Marmite	13
Marmite, spread on bread	540
Marrow, boiled	577
Marrowfat peas, processed	583
Mars bar	168
Mashed potatoes	589
Matzos	21
Mayonnaise	486
Mayonnaise, reduced calorie	487
Meat and vegetable pie, shortcrust pastry top and bottom	319
Meat loaf	320
Meat paste, spread on bread	541
Melon, cantaloupe	245
Melon, honeydew	246
Melon, water	247
Meringue, filled with dairy cream	106
Mild beer, draught	626
Milk chocolate	166
Milk jelly, made up with semi-skimmed milk	422
Milk jelly, made up with skimmed milk	423
Milk jelly, made up with whole milk	424
Milk shake, powder	14
Milk, cows, semi-skimmed, as drink	340
Milk, cows, semi-skimmed, in tea/coffee	345
Milk, cows, semi-skimmed, on cereals	350
Milk, cows, skimmed, as drink	341
Milk, cows, skimmed, in tea/coffee	346
Milk, cows, skimmed, on cereals	351
Milk, cows, whole, as drink	342
Milk, cows, whole, in tea/coffee	347
Milk, cows, whole, on cereals	352
Milk, goats, as drink	343
Milk, goats, in tea/coffee	348
Milk, goats, on cereals	353
Milk, soya, as drink	344
Milk, soya, in tea/coffee	349
Milk, soya, on cereals	354
Mince pie	107
Minced beef, stewed with onion	321
Minced beef, stewed with onion, made with lean mince	322

Minced beef, stewed with vegetables	323
Minced beef, stewed with vegetables, made with lean mince	324
Mixed vegetables, boiled	604
Moussaka	325
Muesli	62
Muesli, with extra fruit	63
Muesli, with no added sugar	64
Muesli/nut low fat yogurt	617
Muffins, bran	108
Muffins, plain	109
Mushroom quiche, made with white flour pastry	160
Mushroom quiche, made with wholemeal flour pastry	161
Mushroom soup, cream of, canned	533
Mushrooms, fried	578
Mushrooms, raw, sliced	464
Mustard and cress	465
Mutton biriani	326
Naan	41
Nectarine	248
Non-dairy ice cream, flavoured	414
Non-dairy ice cream, reduced calorie	415
Non-dairy ice cream, vanilla	413
Nut/muesli low fat yogurt	617
Oatcakes	22
Okra curry	605
Olives, black, with stones	511
Omelette	187
Onion and liver stew	369
Onion sauce, made with semi-skimmed milk	488
Onion sauce, made with skimmed milk	489
Onion sauce, made with whole milk	490
Onion, raw, rings	466
Onions, fried	579
Onions, spring	467
Orange	249
Orange juice, canned, sweetened	525
Orange juice, canned, unsweetened	526
Orange juice, freshly squeezed	527
Orange squash, concentrated, to make up approx. 1/3 pt	528
'Organic' whole milk yogurt	622
Ovaltine, powder	15
Oxtail soup, canned	534
Pakoras	606
Pale ale beer, bottled	629

Pancakes, savoury, made with skimmed milk	2
Pancakes, sweet, made with skimmed milk	425
Pancakes, savoury, made with whole milk	3
Pancakes, sweet, made with whole milk	426
Papadums, fried	42
Paratha	43
Parmesan cheese	140
Parsley, chopped	468
Parsnips, boiled	580
Passion fruit	250
Paste, fish, spread on bread	539
Paste, meat, spread on bread	541
Pâté, liver	370
Pea and potato curry	442
Pea curry	441
Peach	251
Peaches, canned in syrup	265
Peanut butter, spread on bread	542
Peanuts	362
Peanuts, roasted and salted	512
Pear	252
Pears, canned in syrup	266
Peas, frozen, boiled	581
Peas, garden, canned	582
Peas, marrowfat, processed	583
Peppermints	557
Peppers, sweet, raw, sliced	469
Piccalilli	170
Pikelets, toasted	1
Pilau rice	447
Pilchards, in tomato sauce, canned	213
Pineapple	253
Pineapple juice, canned	529
Pineapple, canned in syrup	267
Pitta bread, made with white flour	44
Pitta bread, made with wholemeal flour	45
Pizza	155
Plaice, fillet, in crumbs, fried	215
Plaice, fillets, steamed	214
Plain chocolate	167
Plain digestive biscuits	26
Plain low fat yogurt	618
Plain muffins	109
Plain soufflé	190
Plain whole milk yogurt	623
Plantain, green, boiled	584
Plantain, ripe, fried	585
Plums, dessert	254
Poached egg	184
Popcorn, candied	558

Pork chops, loin, grilled, lean	298
Pork chops, loin, grilled, lean and fat	299
Pork joints, leg, roast, lean	300
Pork joints, leg, roast, lean and fat	301
Pork pie	327
Pork rashers, belly, grilled	302
Pork sausages, grilled	332
Porridge, made with milk	65
Porridge, made with milk and water	66
Porridge, made with water	67
Port wine, fortified	638
Potato and pea curry	442
Potato curry	607
Potato salad	470
Potatoes, boiled	586
Potatoes, chips	587
Potatoes, jacket	588
Potatoes, mashed	589
Potatoes, roast	590
Potatoes, sweet, boiled	591
Prawns, peeled	216
Processed cheese	141
Prunes	274
Puffed wheat	68
Queen of puddings	427
Quiche, cheese, made with white flour pastry	156
Quiche, cheese, made with wholemeal flour pastry	157
Quiche Lorraine, made with white flour pastry	158
Quiche Lorraine, made with wholemeal flour pastry	159
Quiche, mushroom, made with white flour pastry	160
Quiche, mushroom, made with wholemeal flour pastry	161
Radishes	471
Raisins	275
Raspberries	255
Raspberries, canned in syrup	268
Ratatouille	608
Ravioli, in tomato sauce, canned	328
Ready Brek	69
Red cabbage, raw, shredded	454
Red cabbage, sweet and sour	609
Red kidney beans, boiled	567
Red salmon, canned, skin and bone removed	219
Red wine	633

Reduced calorie mayonnaise 487
Reduced calorie salad cream 492
Reduced fat cheddar-type cheese 121
Reduced fat cheshire-type cheese 123
Reduced fat cottage cheese 125
Reduced fat edam-type cheese 131
Ribena, concentrated, to make up
approx. 1/3 pt 530
Rice Krispies 70
Rice pudding, made with semi-skimmed
milk 428
Rice pudding, made with skimmed milk 429
Rice pudding, made with whole milk 430
Ricicles 71
Ricotta cheese 142
Ring doughnuts 92
Ring doughnuts, iced 93
Risotto 448
Roast potatoes 590
Roasted and salted peanuts 512
Rock cake 110
Rock salmon, in batter, fried 217
Roe cod, in crumbs, fried 207
Roquefort cheese 143
Rosé wine 634
Rump beef steaks, grilled, lean 290
Rump beef steaks, grilled, lean and fat 291
Runner beans, boiled 568
Rye crispbread 20

Sage and onion 549
Sage Derby cheese 144
Salad cream 491
Salad cream, reduced calorie 492
Salami 329
Salmon, cutlet, steamed 218
Salmon, red, canned, skin and bone
removed 219
Salmon, smoked 220
Samosa, meat filling 330
Samosa, vegetable filling 610
Sandwich biscuits, e.g. Bourbon,
custard creams 28
Sardines, canned in oil, drained 221
Sardines, canned in tomato sauce 222
Sausage roll, made with flaky pastry 333
Sausages, beef, grilled 331
Sausages, pork, grilled 332
Saveloy 334
Savoury rice 449
Scampi, fried 223
Scones, cheese, made with white flour 513

Scones, fruit, made with white flour 514
Scones, fruit, made with wholemeal
flour 516
Scones, plain, made with white flour 515
Scones, plain, made with wholemeal
flour 517
Scotch egg 188
Scotch pancake 111
Scrambled egg 189
Semi-skimmed cows milk, as drink 340
Semi-skimmed cows milk, in tea/coffee 345
Semi-skimmed cows milk, on cereal 350
Semi-sweet biscuits, e.g. Marie, rich tea 29
Sesame seeds 363
Sevyiaan 431
Shedded Wheat 72
Shepherds pie 335
Sherry wine, fortified, dry 639
Sherry wine, fortified, medium 640
Sherry wine, fortified, sweet 641
Short-sweet biscuits, e.g. Lincoln,
shortcake 30
Shortbread 31
Shreddies 73
Silverside beef joints, boiled, lean 286
Silverside beef joints, boiled, lean
and fat 287
Single cream, in drinks/soups 177
Single cream, on puddings 180
Skimmed condensed milk, sweetened,
on puddings 355
Skimmed cows milk, as drink 341
Skimmed cows milk, in tea/coffee 346
Skimmed cows milk, on cereal 351
Smoked salmon 220
Soufflé, plain 190
Soya cheese 145
Soya milk, as drink 344
Soya milk, in tea/coffee 349
Soya milk, on cereal 354
Soya yogurt 619
Spaghetti, canned in Bolognese sauce 376
Spaghetti, canned in tomato sauce 377
Spaghetti, white, boiled 374
Spaghetti, wholemeal, boiled 375
Special K 74
Spinach, boiled 592
Sponge cake, with butter icing filling 113
Sponge cake, with jam filling 112
Sponge pudding, made with dried fruit 432
Sponge pudding, made with jam or
treacle 433

Spring greens, boiled	593	Topside beef joints, roast, lean and fat	289
Spring onions	467	Treacle tart	434
Steak and kidney pie, with flaky pastry top	336	Trifle, topped with dairy cream	435
		Trifle, topped with Dream Topping	436
Steak and kidney pie, with flaky pastry top and bottom	337	Tripe and onions	373
		Tuna, canned in oil	224
Steak and kidney pudding	338	Turkey, roast, dark meat	385
Stew, beef	310	Turkey, roast, light meat	386
Stew, Irish	316	Turkey, roast, meat and skin	387
Stew, liver and onion	369	Turnips, boiled	598
Stewed apple	395		
Stewed apple, stewed with sugar	396	Vanilla slice	116
Stilton cheese, blue	146	Vegetable curry	611
Stilton cheese, white	147	Vegetable salad, canned	473
Stone fruit jam, e.g. apricot, damson	545	Vegetable soup, canned	536
Stout beer, bottled	630	Vermouth, dry	642
Strawberries	256	Vermouth, sweet	643
Strawberries, canned in syrup	269	Very low fat spread, on bread/roll	195
Streaky bacon rashers, grilled	283	Very low fat spread, on crackers, etc.	200
Sugar Puffs	75	Very low fat spread, with jacket potato	205
Sugar, brown	551		
Sugar, white	552	Wafer biscuits, filled	32
Sultana Bran	76	Walnuts	364
Sultanas	276	Water biscuits	23
Swede, boiled	594	Water melon	247
Sweet and sour red cabbage	609	Watercress	474
Sweet cider	632	Weetabix	77
Sweet peppers, raw, sliced	469	Weetaflakes	78
Sweet pickle, e.g. Branston, Pan Yan	171	Welsh rarebit, made with white bread	162
Sweet potatoes, boiled	591	Welsh rarebit, made with wholemeal bread	163
Sweet vermouth	643		
Sweetcorn kernels, canned	595	Wensleydale cheese	148
Swiss roll	114	Wheatgerm	79
Swiss roll, chocolate mini roll	115	Whipping cream, in drinks/soups	178
		Whipping cream, on puddings	181
Tangerine	257	White bread, from large loaf, medium sliced	34
Taramasalata	231		
Tea, infused	16	White cabbage, raw, shredded	455
Toad in the hole	339	White grapes	241
Toasted crumpets/pikelets	1	White rice, boiled	445
Toffees	559	White sauce, savoury, made with semi-skimmed milk	495
Tofu	443		
Tomato chutney	172	White sauce, savoury, made with skimmed milk	496
Tomato juice, canned	531		
Tomato ketchup	493	White sauce, savoury, made with whole milk	497
Tomato sauce	494		
Tomato soup, cream of, canned	535	White sauce, sweet, made with semi-skimmed milk	505
Tomato, raw	472		
Tomatoes, canned	596	White sauce, sweet, made with skimmed milk	506
Tomatoes, fried	597		
Tongue, canned	372	White sauce, sweet, made with whole milk	507
Topside beef joints, roast, lean	288		

White spaghetti, boiled 374
White Stilton cheese 147
White sugar 552
White wine, dry 635
White wine, sparkling 636
White wine, sweet 637
Whole condensed milk, sweetened, on
  puddings 356
Whole cows milk, as drink 342
Whole cows milk, in tea/coffee 347
Whole cows milk, on cereal 352
Whole evaporated milk, on puddings 357
Whole milk yogurt, fruit 620
Whole milk yogurt, goats milk 621
Whole milk yogurt, 'organic' 622
Whole milk yogurt, plain 623
Wholemeal bread, from large loaf,
  medium sliced 35
Wholemeal crackers 18
Wholemeal spaghetti, boiled 375
Wine, fortified, Port 638
Wine, fortified, Sherry, dry 639
Wine, fortified, Sherry, medium 640
Wine, fortified, Sherry, sweet 641

Wine, red 633
Wine, rosé 634
Wine, white, dry 635
Wine, white, sparkling 636
Wine, white, sweet 637

Yam, boiled 599
Yogurt, Greek, cows milk 612
Yogurt, Greek, ewes milk 613
Yogurt, low calorie, assorted flavours 614
Yogurt, low fat, flavoured 615
Yogurt, low fat, fruit 616
Yogurt, low fat, muesli/nut 617
Yogurt, low fat, plain 618
Yogurt, soya 619
Yogurt, whole milk, fruit 620
Yogurt, whole milk, goats milk 621
Yogurt, whole milk, 'organic' 622
Yogurt, whole milk, plain 623
Yorkshire puddings, made with skimmed
  milk 4
Yorkshire puddings, made with whole milk 5